Kleene-Schützenberger and Büchi Theorems for Weighted Timed Automata

Der Fakultät für Mathematik und Informatik
der Universität Leipzig
angenommene

DISSERTATION

zur Erlangung des akademischen Grades

Doctor rerum naturalium

(Dr. rer. nat.)

im Fachgebiet

Informatik

vorgelegt von

MSc Karin Quaas

geboren am 14. Oktober 1979 in Leipzig.

Die Annahme der Dissertation haben empfohlen:

1. Prof. Dr. Manfred Droste (Leipzig)
2. Dr. habil. Patrica Bouyer-Decitre (Cachan)

Die Verleihung des akademischen Grades erfolgt mit Bestehen
der Verteidigung am 24.03.2010 mit dem Gesamtprädikat *summa cum laude*.

Bibliografische Information der Deutschen Nationalbibliothek

Die Deutsche Nationalbibliothek verzeichnet diese Publikation in der Deutschen Nationalbibliografie; detaillierte bibliografische Daten sind im Internet über http://dnb.d-nb.de abrufbar.

ISBN 978-3-8325-2500-2

Logos Verlag Berlin GmbH
Comeniushof, Gubener Str. 47,
10243 Berlin
Tel.: +49 (0)30 42 85 10 90
Fax: +49 (0)30 42 85 10 92
INTERNET: http://www.logos-verlag.de

Acknowledgments

First of all I would like to thank Manfred Droste for his constant support and encouragement during the last three years, for giving me the possibility to research in his group, and for sharing his expertise.

Furthermore, I would like to thank Patricia Bouyer-Decitre for reviewing this thesis.

I am grateful for the financial support of the Deutsche Forschungsgemeinschaft through the Graduiertenkolleg "Wissensrepräsentation" and the Landesgraduiertenstipendum Sachsen.

Finally I would like to thank my friends and family for their support.

Contents

1 Introduction **1**

2 Preliminaries **5**
- 2.1 Timed Automata 5
- 2.2 Semirings and Formal Power Series 9
- 2.3 Weighted Finite Automata 11

3 Weighted Timed Automata **15**
- 3.1 Relation To Other Automata Models 18
- 3.2 Closure Properties of Recognizable Timed Series 20

4 A Kleene-Schützenberger Theorem for Weighted Timed Automata **27**
- 4.1 Introduction 27
- 4.2 Clock Series 28
- 4.3 From Rationality to Recognizability 31
- 4.4 From Recognizability to Rationality 42
- 4.5 From Clock Series to Timed Series 46
- 4.6 Conclusion 48

5 A Büchi Theorem for Weighted Timed Automata **51**
- 5.1 Introduction 51
- 5.2 Weighted Relative Distance Logic 52
- 5.3 From Definability to Recognizability 56
- 5.4 From Recognizability to Definability 66
- 5.5 Generalizations to Arbitrary Semirings 69
- 5.6 Conclusion 78

6 Supports and Timed Cut Languages **81**
- 6.1 Introduction 81
- 6.2 Recognizability of Supports of Recognizable Timed Series 82
- 6.3 The Empty Support Problem for Recognizable Timed Series over Fields . 88
- 6.4 Timed Cut Languages 92
- 6.5 Conclusions and Further Research 101

7 Conclusion and Future Work **103**

Bibliography **107**

Index **117**

1 Introduction

Since its introduction in the seminal paper by Alur and Dill [6], *timed automata* have been a thoroughly investigated model for the specification and analysis of real-time systems. Basically, a timed automaton is a finite automaton equipped with a finite set of *clocks* that measure the time and may be used to put restrictions on when it is allowed to take a transition. The behaviour of a timed automaton is captured by the notion of *timed languages*, an extension of classical formal languages, where each symbol in a word is paired with the time of its occurrence. In their paper, Alur and Dill develop a theory of timed automata comprising timed extensions of classical decision problems and closure properties (see also [10] for a survey). Besides this, much research work has been done to generalize other results known from the classical theory of formal languages to the timed setting. For instance, there are various approaches to provide alternative characterizations of timed automata and their behaviours. For instance, the famous Büchi theorem [36] about the coincidence of recognizable languages and languages definable in monadic second-order logic (MSO) is extended to timed languages by Wilke [107]. Also, several generalizations of the Kleene theorem [76] are proposed [12, 14, 13, 28, 29, 59, 60]. But first and foremost, there is much research concerning the theoretical foundations for the development of efficient verification and model checking tools, e.g. UPPAAL, KRONOS or HYTECH [31, 79, 71]. Tools like these have successfully been used for solving industrially relevant problems, like, for instance, the detection and correction of an implementation error in an audio and video protocol used by BANG & OLUFSEN [68], or the development of a clock synchronization algorithm that is currently used in a wireless sensor network developed by the Dutch company CHESS [69], just to mention two of them.

The success of timed automata is certainly due to the decidability of the *reachability problem*, i.e., the problem to decide, whether in a given timed automaton one can reach a given state starting from an initial state [6]. The reachability problem is of high importance. For instance, in the field of verification one is often interested in checking whether a given "bad" state can never be reached. Furthermore, some important decision problems (e.g. the emptiness problem) can be reduced to it. However, in a model for *timed* systems it is a natural idea not only to ask whether a state can be reached, but also to ask for the *fastest* way to reach it. This first has been considered in a paper by Niebert et al. [89], and the practical relevance e.g. for planning and scheduling problems has been discussed in several papers, e.g. [2, 3, 1]. Certainly, it is not only interesting to find the *fastest* but also, in a more general sense, the *cheapest* way to a state. This leads directly to the model of *weighted timed automata* (also known as *priced timed automata*), which has been introduced in 2001 by Alur et al. [8] and Behrmann et al. [17]. A weighted timed automaton is a timed automaton extended with a cost function that assigns a natural number to both the transitions and states of the timed automaton. The natural number assigned to a transition is interpreted as the cost for taking this transition. The natural number assigned to a state denotes the cost rate for the time spent in that state. In this way, weighted timed automata can be used to model the *continuous* consumption of resources. The costs do not influence the behaviour of the automaton, but may be used as a measure of performance. For instance, as mentioned above, one may be interested in finding the cost optimal way to a given state. This problem, called the *optimal cost reachability problem*, is known to be computable [8, 9, 17, 22]. The model of weighted timed automata has inspired researchers of the real-time community to much other research work of practical relevance. For instance, there are decidability results on *weighted timed games*, which are useful for synthesizing controllers [4, 33, 21, 23, 27]. The model also led to weighted extensions of temporal logics with cost-contrained modalities, known as WCTL and WMTL, and the corresponding model checking problems are considered e.g. in [32, 21, 23, 34, 26]. The model of *multi-priced timed automata*, where more than one cost variable is permitted, allows for very promising applications e.g. in the modelling of embedded systems, where often more than one resource (e.g. bandwidth, memory, energy) must be restricted. Results on weighted timed games, optimal reachability and model checking using multiple cost variables can be found for instance in [80, 24, 81, 25, 65].

The motivation behind most of this work on weighted timed automata is to provide the real-time community with decidability results and algorithms for building efficient verification tools. However, not much is known so far about the model of weighted timed automata itself, i.e., about closure properties or characterizations using algebraical and logical formalisms. In fact, Asarin [11] points out a similar situation for the model of timed automata. He compares the state of knowledge about timed automata with a building having a bright and well constructed roof, but lacking of the necessary basic foundations. In this sense, we want to consider weighted timed automata not only as an interesting tool used for verification purposes, but we want to investigate theoretical properties of the model itself. A theoretical perspective on the model may also give rise to new practically relevant ideas.

In the following, we summarize the content of the thesis. First of all, in Chapter 3, we define weighted timed automata *over a semiring* in the same manner as it is done for classical weighted finite automata [99, 77, 19, 50]. In this way, we are not bound to a fixed set of costs or weights, nor are we restricted to use the operations of addition and infimum for computing the weight of a word. Secondly, we do not restrict the cost functions for the locations to be linear (as previously done in [8, 17]). Instead, we consider weighted timed automata with respect to an arbitrary family of functions mapping positive reals to elements in the semiring. The cost for staying in a location is defined by a function of this family. Doing so, we obtain a flexible model of weighted timed automata. Not only most of the models of weighted timed automata proposed in the literature so far are special cases of our model, but our definition gives also rise to interesting new instances. By using semirings, we build a bridge to the theory of weighted finite automata. This stimulates the idea to extend well known classical notions from the theory of weighted finite automata to the timed setting. In our model, the behaviour of a weighted timed automaton is defined by the new notion of *timed series*, i.e., timed extensions of formal power series, which map each timed word to a coefficent in the semiring. In fact, the behaviour of a weighted timed automaton has never been investigated before with respect to the timed language that is recognized. This new definition of weighted timed automata was first published in [53].

In Chapters 4 and 5, we aim to establish alternative characterizations of the behaviours of weighted timed automata. We start with a generalization of the fundamental Kleene theorem for finite automata over words [76], stating that the class of recognizable languages equals the class of languages that can be defined by *rational expressions*. This theorem was generalized to formal power series and weighted finite automata by Schützenberger [101]. Also, several approaches [12, 14, 13, 28, 29, 59, 60] were done to obtain such a type of theorem for timed automata. For our result, we use the latest approach of Bouyer and Petit [29] and combine it with the results of Schützenberger and new techniques. As it is not possible to define a natural concatenation operation over the class of timed words, we introduce the notion of *clock series*, an extension of *clock words* introduced by Bouyer and Petit, and define weighted extensions of the classical rational operations sum, concatenation and (finite) Kleene star iteration over the set of clock series. We then show that recognizable clock series are closed under these operations, from which we can conclude that every rational clock series is recognizable. For the other direction, i.e., that each recognizable clock series is rational, we use an extension of a classical approach proposed by Brzozowski [35] and also used by Bouyer and Petit [29]. Lastly, we show how we can obtain a Kleene theorem also for timed series. An extended abstract of the results presented in this chapter appeared in [53].

In Chapter 5, we lift the classical theorem on the equivalence of the expressive power of finite automata and MSO logic to the weighted timed setting. A theorem of this type was first given by Büchi [36] and has since then been generalized to many other classes of automata. For timed automata, Wilke [107] introduced an MSO logic, called the *relative distance logic*, and showed that every timed language that can be defined by sentences in this logic can be recognized by a timed automaton, and vice versa. A corresponding result for the class of weighted finite automata was recently presented by Droste and Gastin [46, 49]. The authors define a weighted MSO logic, where atomic formulas may additionally comprise elements of the semiring. The operators of the logic are given a weighted semantics. In this way, a formula of this logic defines a formal power series. Droste and Gastin show that a fragment of this logic is expressively equivalent to the class of weighted finite automata. For presenting a Büchi theorem for the class of weighted timed automata, we will combine the weighted MSO logic of Droste and Gastin, Wilke's relative distance logic and a new kind of weighted formulas to express the weights that arise while being in a state. Then we aim to show the equivalence of recognizable and definable timed series. Similarly to the untimed setting, we need to define a suitable fragment of our weighted timed MSO logic. Since the definitions and constructions are technical, mainly due to the weights that arise while letting time elapse, we do this stepwisely and start by restricting our considerations to idempotent and commutative semirings. Later on, we generalize the results to arbitrary semirings. However, neglecting the idempotence of the semiring goes at the expense of the timed languages we are allowed to use. More detailed, we need to restrict the universal first-order quantifier to formulas which define timed languages of bounded variability, a notion introduced by Wilke [107]. The reason for this is mainly due to the non-determinizability of timed automata [6]. Also, if we skip the restriction on the semiring being commutative, we are faced with some problems which may be solved by restricting the weight functions or excluding empty timed words. Some of the constructions of this chapter were presented for a subclass of weighted timed automata, namely weighted *event-recording automata* [7], in an extended abstract [94].

In Chapter 6, we turn towards the *support* and *cut languages* of timed series. Both notions are adopted from the theory of weighted finite automata and formal power series. The support of a timed series is the timed language containing all timed words which are not mapped to the zero element of the semiring. Supports have been investigated extensively in the theory of weighted finite automata and formal power series [19, 105, 98, 75, 74]. Here, we are interested in transferring some of the results to the timed setting. For instance, using a well known result that can already be found in [19], we can show that the support of the behaviour of each weighted timed automaton as introduced by Alur et al. [9] is recognizable by a timed automaton. Using a weighted variant of the classical region automaton construction for timed automata [6], we present a procedure to decide whether the support of a weighted timed automaton over a field (as e.g. the real numbers with addition and multiplication) and the family of linear functions is empty or not. Note that this is a weighted version of the classical emptiness problem. As a consequence of this result, we obtain the decidability of the equivalence problem for weighted timed automata over the reals and linear functions. This is a remarkable result, as for unweighted timed automata, the corresponding problem is undecidable [6]. The result also raises the question whether one can develop new algorithms for this kind of weighted timed automata. Finally, we investigate the recognizability of *timed cut languages*, i.e., sets of timed words which are assigned a value smaller than or greater than a given value from the semiring. Parts of this chapter are published in [93].

2 Preliminaries

Let $\mathbb{N}$, $\mathbb{Z}$, $\mathbb{Q}$ and $\mathbb{R}$ denote the set of natural, integer, rational and real numbers, respectively. For $K \in \{\mathbb{Z}, \mathbb{Q}, \mathbb{R}\}$, we use $K_{\geq 0}$ to denote the respective positive numbers. We use Σ and Γ to denote finite alphabets. We use the symbol $\uplus$ to denote a disjoint union of finite sets.

2.1 Timed Automata

We consider *timed automata*, a basic and natural model to represent the behaviour of real-time systems, which has been introduced in a seminal paper by Alur and Dill [6]. Intuitively, a timed automaton is a finite automaton equipped with a finite set of clocks ranging over $\mathbb{R}_{\geq 0}$ (see Fig. 2.1). While being in a state (called *location* in a timed automaton), time elapses and the values of the clocks increase. The model allows the transitions (called *edges*) to be labelled with boolean combinations of atomic formulas which compare clock values with natural constants. Edges may only be taken if the formulas are satisfied. Furthermore, clocks may be reset to zero at the edges independently of each other. In the following, we introduce some preliminary notions and explain the model of timed automata. Then we give some decidability results that will be used later.

A *timed word* over Σ is a finite sequence $(a_1, t_1)...(a_k, t_k) \in (\Sigma \times \mathbb{R}_{\geq 0})^*$ such that the sequence $\bar{t} = t_1...t_k$ of timestamps is non-decreasing. Sometimes we denote a timed word as above by $(\bar{a}, \bar{t})$, where $a \in \Sigma^*$. We write $T\Sigma^*$ for the set of timed words over Σ. The empty timed word is denoted by ε and we define $T\Sigma^+ = T\Sigma^* \backslash \{\varepsilon\}$. A set $L \subseteq T\Sigma^*$ is called a *timed language*. We say that a timed word is *strictly monotonic* if its sequence of timestamps is strictly monotonically increasing (i.e., we rule out zero time delays). We use $T_s\Sigma^*$ to denote the set of strictly monotonic timed words. A timed language $L \subseteq T_s\Sigma^*$ is called strictly monotonic. Given a timed word w as above we let the domain $\mathsf{dom}(w)$ be $\{1, ..., k\}$ and define the length $|w|$ of w to be k. For $a \in \Sigma$, we let $|w|_a$ be

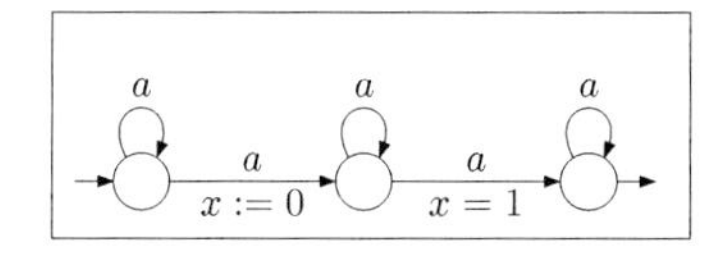

Figure 2.1: A timed automaton with a single clock

the number of occurrences of a in w. Let $\pi : \Sigma \to \Gamma$ be a mapping. The *renaming* $\pi(w)$ of a timed word $w \in T\Sigma^*$ is the timed word $w' \in T\Gamma^*$ of the form $(a'_1, t'_1)...(a'_k, t'_k)$ such that $a'_i = \pi(a_i)$ and $t'_i = t_i$ for all $i \in \mathsf{dom}(w)$.

Let $\mathcal{C}$ be a finite set of *clock variables* ranging over $\mathbb{R}_{\geq 0}$. We define *clock constraints* ϕ over $\mathcal{C}$ to be conjunctions of formulas of the form $x \sim c$, where $c \in \mathbb{N}$, $x \in \mathcal{C}$ and $\sim \in \{<, \leq, =, \geq, >\}$. Let $\Phi(\mathcal{C})$ be the set of all clock constraints over $\mathcal{C}$. A *clock valuation* $\nu : \mathcal{C} \to \mathbb{R}_{\geq 0}$ is a function that assigns a value to each clock variable. A clock valuation ν satisfies a clock constraint ϕ, written $\nu \models \phi$, if ϕ evaluates to true according to the values given by ν. For $\delta \in \mathbb{R}_{\geq 0}$ and $\lambda \subseteq \mathcal{C}$, we define the clock valuation $\nu + \delta$ to be $(\nu+\delta)(x) = \nu(x)+\delta$ for each $x \in \mathcal{C}$ and the clock valuation $\nu[\lambda := 0]$ by $(\nu[\lambda := 0])(x) = 0$ if $x \in \lambda$ and $(\nu[\lambda := 0])(x) = \nu(x)$ otherwise.

A *timed automaton* over Σ is a tuple $\mathcal{A} = (\mathcal{L}, \mathcal{L}_0, \mathcal{L}_f, \mathcal{C}, E)$, where

- $\mathcal{L}$ is a finite set of locations,
- $\mathcal{L}_0 \subseteq \mathcal{L}$ is a set of initial locations,
- $\mathcal{L}_f \subseteq \mathcal{L}$ is a set of final locations,
- $\mathcal{C}$ is a finite set of clock variables,
- $E \subseteq \mathcal{L} \times \Sigma \times \Phi(\mathcal{C}) \times 2^{\mathcal{C}} \times \mathcal{L}$ is a finite set of edges. An edge $(l, a, \phi, \lambda, l')$ allows a jump from location l to location l' if a is read, provided that for the current valuation ν we have $\nu \models \phi$. After the edge has been executed, the new valuation is $\nu[\lambda := 0]$. We use $\mathsf{source}(e)$ to denote the source location l and $\mathsf{lab}(e)$ to denote the label a of an edge e.

A state in a timed automaton is a pair $(l, \nu) \in \mathcal{L} \times (\mathbb{R}_{\geq 0})^{\mathcal{C}}$. Between states there are two kinds of transitions. A *timed transition* is of the form $(l, \nu) \xrightarrow{\delta} (l, \nu + \delta)$ for some $\delta \in \mathbb{R}_{\geq 0}$ and represents the elapse of δ time units in l. A *discrete transition* is of the form $(l, \nu) \xrightarrow{e} (l', \nu')$ for some $e = (l, a, \phi, \lambda, l') \in E$ such that $\nu \models \phi$ and $\nu' = \nu[\lambda := 0]$. We write $(l, \nu) \xrightarrow{\delta}\xrightarrow{e} (l', \nu')$ to denote that there is a state (l'', ν'') such that $(l, \nu) \xrightarrow{\delta} (l'', \nu'')$ and $(l'', \nu'') \xrightarrow{e} (l', \nu')$. We call such a timed transition followed by a discrete transition a *transition*. A *run* of $\mathcal{A}$ on a timed word w is a finite sequence $(l_0, \nu_0) \xrightarrow{\delta_1}\xrightarrow{e_1} (l_1, \nu_1) \xrightarrow{\delta_2}\xrightarrow{e_2} \dots \xrightarrow{\delta_{|w|}}\xrightarrow{e_{|w|}} (l_{|w|}, \nu_{|w|})$ of transitions, where $\nu_0 = 0^{\mathcal{C}}$. We say that r is *successful* if $l_0 \in \mathcal{L}_0$ and $l_{|w|} \in \mathcal{L}_f$. The timed language $L(\mathcal{A})$ recognized by a timed automaton $\mathcal{A}$ is defined by $L(\mathcal{A}) = \{w \in T\Sigma^* : \text{there is a successful run of } \mathcal{A} \text{ on } w\}$. A timed language $L \subseteq T\Sigma^*$ is said to be *TA-recognizable* over Σ if there is a timed automaton $\mathcal{A}$ over Σ such that $L(\mathcal{A}) = L$. If Σ is clear from the context, we may omit it.

Example 2.1. Let $\Sigma = \{a\}$ and consider the timed language $L_{\text{nd}} = \{w \in T\Sigma^* : \exists i, j \in \mathsf{dom}(w). t_j - t_i = 1\}$. This timed language can be recognized by the single-clock timed automaton shown in Fig. 2.1. The clock constraint $x = 1$ and the clock reset ensure that the edge between the middle and the right location is only taken if there is some a-event that happened exactly one time unit ago.

A timed automaton $\mathcal{A}$ is *deterministic* if $|\mathcal{L}_0| = 1$, and whenever $(l, a, \phi_1, \lambda_1, l_1)$ and $(l, a, \phi_2, \lambda_2, l_2)$ are two different edges in $\mathcal{A}$, then for all clock valuations ν we have $\nu \not\models \phi_1 \wedge \phi_2$. $\mathcal{A}$ is *unambiguous* if for every timed word $w \in L(\mathcal{A})$ there is exactly one successful run of $\mathcal{A}$ on w. A timed language $L \subseteq T\Sigma^*$ is *deterministically TA-recognizable* (*unambiguously TA-recognizable*, respectively) if there is a deterministic (unambiguous, respectively) timed automaton $\mathcal{A}$ over Σ such that $L(\mathcal{A}) = L$.

Proposition 2.2 ([6, 10]). *The class of TA-recognizable timed languages is closed under union, intersection, renaming and inverse renaming.*

The proof that intersection preserves TA-recognizability of timed languages involves the usual product construction known from the classical theory of formal languages. Contrary to the classical case, both the class of deterministically TA-recognizable and the class of unambiguously TA-recognizable timed languages form strict subclasses of TA-recognizable timed languages.

Example 2.3. The timed language L_{nd} from Ex. 2.1 can neither be recognized by a deterministic timed automaton nor is its complement TA-recognizable [6].

Example 2.4. The timed language L_{amb} containing all timed words w satisfying the following conditions:

1. $t_i < 2$ for all $i \in \mathsf{dom}(w)$,
2. $t_i = 1$ for some $i \in \mathsf{dom}(w)$,
3. there are $1 \leq i < j \leq \mathsf{dom}(w)$ such that $t_j - t_i = 1$.

cannot be recognized by an unambiguous timed automaton [106].

The class of unambiguously TA-recognizable timed languages is closed under union and intersection, but it is not known whether it is closed under complement. However, it is decidable for a TA-recognizable timed language L whether it is unambiguously TA-recognizable [106].

Theorem 2.5 ([6]). *The emptiness problem for timed automata is decidable.*

The solution to the emptiness problem involves the construction of a finite quotient of the infinite state space induced by a timed automaton, called the *region automaton* [6]. In the following, we recall the definition of the region automaton as it will be needed later.

Let $\delta \in \mathbb{R}_{\geq 0}$. We use $\langle \delta \rangle$ to denote the fractional part of δ, and $\lfloor \delta \rfloor$ to denote the integral part of δ (also known as the *floor* of δ), respectively. For each clock variable $x \in \mathcal{C}$, we use c_x to denote the largest integer c that x is compared with in some clock constraint of an edge. We define an equivalence relation over the set of all clock valuations for $\mathcal{C}$, called the *region equivalence* [6], as follows. Two clock valuations $\nu, \nu' \in (\mathbb{R}_{\geq 0})^{\mathcal{C}}$ are region equivalent, written $\nu \cong \nu'$, if the following three conditions are satisfied.

1. For all clock variables $x \in \mathcal{C}$, either $\lfloor \nu(x) \rfloor = \lfloor \nu'(x) \rfloor$, or both $\nu(x)$ and $\nu'(x)$ exceed c_x.
2. For all clock variables $x, x' \in \mathcal{C}$ with $\nu(x) \leq c_x$ and $\nu(x') \leq c_{x'}$, we have $\langle \nu(x) \rangle \leq \langle \nu(x') \rangle$ iff $\langle \nu'(x) \rangle \leq \langle \nu'(x') \rangle$.
3. For all clock variables $x \in \mathcal{C}$ with $\nu(x) \leq c_x$, we have $\langle \nu(x) \rangle = 0$ iff $\langle \nu'(x) \rangle = 0$.

An equivalence class of clock valuations induced by $\cong$ is called a *clock region*. We define the *initial clock region* to be the equivalence class of the clock valuation ν_0 that maps all clock variables to zero. Define $c_{\max} = \max\{c_x : x \in \mathcal{C}\}$. We define the finite alphabet $\mathcal{I} = \{[0,0], (0,1), [1,1], (1,2), ..., (c_{\max} - 1, c_{\max}), [c_{\max}, c_{\max}], (c_{\max}, \infty)\}$ of intervals over $\mathbb{R}_{\geq 0}$. Let $w = (a_1, t_1)...(a_k, t_k)$ be a timed word. We use $\mathsf{abs}(w)$ to denote the unique untimed word $(I_1, a_1)(I_2, a_2)...(I_k, a_k)$ over $\mathcal{I} \times \Sigma$, where for each $i \in \{1, ..., k\}$, I_i is the unique interval in $\mathcal{I}$ with $t_i - t_{i-1} \in I_i$. Given a timed automaton $\mathcal{A} = (\mathcal{L}, \mathcal{L}_0, \mathcal{L}_f, \mathcal{C}, E)$ over Σ, we define the region automaton $R(\mathcal{A}) = (Q, Q_0, Q_f, \Delta)$ to be a finite automaton over $\mathcal{I} \times \Sigma$, where

- $Q = \{(l, r) : l \in \mathcal{L}, r \text{ is a clock region}\}$,
- $Q_0 = \{(l, r) : l \in \mathcal{L}_0, r \text{ is the initial clock region}\}$,
- $Q_f = \{(l, r) : l \in \mathcal{L}_f, r \text{ is a clock region}\}$,
- $\big((l, r), (I, a), (l', r')\big) \in \Delta$ if and only if there are clock valuations $\nu \in r$, $\nu' \in r'$, $e \in E$ with $\mathsf{lab}(e) = a$, and $\delta \in I$ such that $(l, \nu) \xrightarrow{\delta}\xrightarrow{e} (l', \nu')$ is a transition of $\mathcal{A}$. We say that $\big((l, r), (I, a), (l', r')\big) \in \Delta$ stems from e and δ.

The region automaton is finite and bisimulation equivalent to the infinite state-transition system induced by the corresponding timed automaton [39].

The universality problem, and hence also the language inclusion and language equivalence problems for timed automata are undecidable [6]. However, the language inclusion problem becomes decidable if we restrict the number of clock variables to one.

Theorem 2.6 ([91]). *Given two timed automata $\mathcal{A}$ and $\mathcal{A}'$, where $\mathcal{A}$ has at most one clock, it is decidable whether $L(\mathcal{A}') \subseteq L(\mathcal{A})$ and whether $L(\mathcal{A}) = T\Sigma^*$.*

2.2 Semirings and Formal Power Series

In this section, we present the mathematical formalisms needed for defining the model of weighted finite automata, i.e., finite automata extended with a function assigning weights to the transitions. In a very general way, these weights come from a semiring. The behaviour of a weighted finite automaton corresponds to a function that assigns to each finite word a coefficient in the semiring, namely its weight. Such a function is called a formal power series.

A monoid $(K, \cdot, 1)$ is a set K together with an associative operation $\cdot$ and a unit element 1 such that $1 \cdot k = k \cdot 1 = k$ for each $k \in K$. A monoid is commutative if additionally $\cdot$ is commutative. A semiring $\mathcal{K}$ is an algebraic structure $(K, +, \cdot, 0, 1)$ where $(K, +, 0)$ is a commutative monoid, $(K, \cdot, 1)$ is a monoid, multiplication distributes over addition, and multiplication is absorbing, i.e., $0 \cdot k = k \cdot 0 = 0$ for each $k \in K$. In the following, we give some examples of important semirings.

- the semiring $(\mathbb{R}, +, \cdot, 0, 1)$ of the real numbers with ordinary addition and multiplication
- the Boolean semiring $(\{0, 1\}, \vee, \wedge, 0, 1)$
- the min-plus-semiring $(\mathbb{R}_{\geq 0} \cup \{\infty\}, \min, +, \infty, 0)$
- the max-plus-semiring $(\mathbb{R}_{\geq 0} \cup \{-\infty\}, \max, +, -\infty, 0)$
- the Viterbi-semiring $([0, 1], \max, \cdot, 0, 1)$
- the min-max-semiring $(\mathbb{R} \cup \{\infty, -\infty\}, \min, \max, \infty, -\infty)$

We say that a semiring is *commutative* if $(K, \cdot, 1)$ is a commutative monoid. If the addition in $\mathcal{K}$ is idempotent, i.e., we have $k + k = k$ for each $k \in K$, then the semiring is called *idempotent*. A semiring is *zero-divisor free* if for all $k, k' \in K$, whenever $k \cdot k' = 0$ then $k = 0$ or $k' = 0$. Similarly, a semiring is *zero-sum free* if for all $k, k' \in K$, whenever $k + k' = 0$ then $k = 0$ and $k' = 0$. We say that a semiring is *positive* if it is both zero-divisor free and zero-sum free. A semiring is a *field* if $(K, +, 0)$ is a group and $(K \backslash \{0\}, \cdot, 1)$ is a commutative group. Let $A \subseteq K$. The *subsemiring generated by* A is the least subset of K which includes A, 0 and 1 and is closed both under $+$ and $\cdot$. If A is finite, this subsemiring is *finitely generated.* A semiring $\mathcal{K}$ is *locally finite* if each finitely generated subsemiring is finite. Notice that this is the case if and only if the monoids $(K, +, 0)$ and $(K, \cdot, 1)$ are locally finite [47]. If $(K, +, 0)$ is a group, then $\mathcal{K}$ is a *ring.* A semiring $\mathcal{K}$ has characteristic zero if there is no $n \in \mathbb{N} \backslash \{0\}$ such that $\underbrace{1 + ... + 1}_{n} = 0$.

A *monoid morphism* between monoids $(K, \cdot, 1)$ and $(K', \cdot', 1')$ is a function $\eta : K \to K'$ satisfying

- $\eta(1) = 1'$,
- $\eta(a \cdot b) = \eta(a) \cdot' \eta(b)$ for all $a, b \in K$.

Accordingly, a *semiring morphism* between two semirings $(K, +, \cdot, 0, 1)$ and $(K', +', \cdot', 0', 1')$ is a function $\eta : K \to K'$ such that η is both a monoid morphism between $(K, +, 0)$ and $(K', +', 0')$ and a monoid morphism between $(K, \cdot, 1)$ and $(K', \cdot', 1')$. Let $A, B \subseteq K$ be two subsets of the semiring $\mathcal{K}$. We say that A and B *commute element-wise* if $a \cdot b = b \cdot a$ for each $a \in A, b \in B$. Let $\mathcal{K}_A$ be the subsemiring of $\mathcal{K}$ generated by A. Notice that each element $k \in K_A$ can be written as a finite sum of finite products of elements in A. From this it follows that whenever A and B commute element-wise, then K_A and K_B commute element-wise.

A mapping $S : \Sigma^* \to K$ is called a *formal power series*, or *series*, for short. For historical reasons, we write (S, w) instead of $S(w)$ for each $w \in \Sigma^*$. The *support* of S, denoted by $\mathsf{supp}(S)$, is defined to be the set $\{w \in \Sigma^* : (S, w) \neq 0\}$. The set of all series over $\mathcal{K}$ and Σ is denoted by $\mathcal{K}\langle\langle\Sigma^*\rangle\rangle$. Let $S, S_1, S_2 \in \mathcal{K}\langle\langle\Sigma^*\rangle\rangle$ and $k \in K$. We define the *sum* $S_1 + S_2$, the *Hadamard product* $S_1 \odot S_2$, the *Cauchy product* $S_1; S_2$, and the scalar products $S \cdot k$ and $k \cdot S$ as follows. For each $w \in \Sigma^*$, we have

$$\begin{aligned}
(S_1 + S_2, w) &= (S_1, w) + (S_2, w), \\
(S_1 \odot S_2, w) &= (S_1, w) \cdot (S_2, w), \\
(S_1; S_2, w) &= \sum_{w_1; w_2 = w} (S_1, w_1) \cdot (S_2, w_2), \\
(S \cdot k, w) &= (S, w) \cdot k, \\
(k \cdot S, w) &= k \cdot (S, w),
\end{aligned}$$

where ; denotes the natural concatenation operation of two finite words. Notice that the first three operations correspond to union, intersection and concatenation, respectively, of formal languages if we let $\mathcal{K}$ be the Boolean semiring. Also note that the sum in the definition of the Cauchy product is finite and hence well-defined in $\mathcal{K}$, since any $w \in \Sigma^*$ has only finitely many decompositions as $w = w_1; w_2$. For $k \in K$, we further define the constant series k, which maps each $w \in \Sigma^*$ to k, and the series $k\varepsilon$ defined by $(k\varepsilon, w) = k$ if $w = \varepsilon$ and $(k\varepsilon, w) = 0$ otherwise. Then, $(\mathcal{K}\langle\langle\Sigma^*\rangle\rangle, +, \odot, 0, 1)$ and $(\mathcal{K}\langle\langle\Sigma^*\rangle\rangle, +, ;, 0, 1_\varepsilon)$ are again semirings. This can be proved by elementary calculations. For $L \subseteq \Sigma^*$, we define the *characteristic series* $1_L : \Sigma^* \to K$ by $(1_L, w) = 1$ if $w \in L$ and $(1_L, w) = 0$ otherwise. Let $\pi : \Sigma \to \Gamma$ be a mapping. The *renaming* of a finite word $a_1 ... a_k$ over Σ is defined to be the finite word $\pi(a_1)...\pi(a_k)$ over Γ. Given $S \in \mathcal{K}\langle\langle\Sigma^*\rangle\rangle$, we define the renaming $\bar{\pi}(S) : \Gamma^* \to K$ by $(\bar{\pi}(S), u) = \sum_{\pi(w)=u} (S, w)$ for all $u \in \Gamma^*$. Notice that, again, the sum in this equation is finite and thus well-defined in $\mathcal{K}$. For a series $S \in \mathcal{K}\langle\langle\Gamma^*\rangle\rangle$, we define the *inverse renaming* $\bar{\pi}^{-1}(S) : \Sigma^* \to K$ by $(\bar{\pi}^{-1}(S), w) = (S, \pi(w))$ for each $w \in \Sigma^*$.

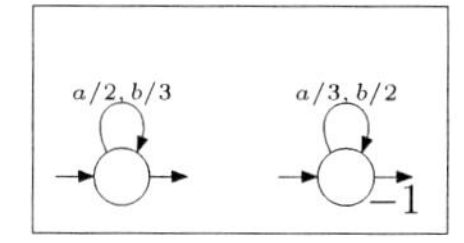

Figure 2.2: A weighted finite automaton over the semiring of the real numbers

2.3 Weighted Finite Automata

Now, we present the model of weighted finite automata, which is the subject of many works [99, 77, 19, 50]. As mentioned before, a weighted finite automaton is a finite automaton extended with a function that assigns weights from a semiring to the transitions. This model may, amongst others, be used to model the consumption of resources. Weighted finite automata are of practical interest e.g. in natural language applications, in particular speech processing [87, 40, 88], and image or video compression [41, 67, 73]. Recently, they have found their way into a branch of verification, namely multi-valued model checking [78]. We fix a semiring $\mathcal{K}$ and an alphabet Σ. A weighted finite automaton $\mathcal{A}$ over $\mathcal{K}$ and Σ is a tuple $\mathcal{A} = (Q, \Delta, \mathsf{in}, \mathsf{out}, \mathsf{wt})$, where

- Q is a finite set of states,
- $\Delta \subseteq Q \times \Sigma \times Q$ is a set of transitions,
- $\mathsf{in} : Q \to K$ is a weight function for entering a state,
- $\mathsf{out} : Q \to K$ is a weight function for leaving a state,
- $\mathsf{wt} : \Delta \to K$ is a weight function for taking a transition.

For representing weighted finite automata graphically, we use the following conventions. An incoming (outgoing, respectively) arrow labeled with a value from K represents the weight for entering (leaving, respectively) the state; we omit the value if it equals 1. The weight for taking a transition is given behind the label of the transition and sometimes omitted if equal to 1.

Example 2.7. In Fig. 2.2, we show a weighted finite automaton $\mathcal{A}$ over the semiring of the real numbers with addition and multiplication, and $\Sigma = \{a, b\}$. We have $(\|\mathcal{A}\|, w) = 2^{|w|_a} \cdot 3^{|w|_b} - 3^{|w|_a} \cdot 2^{|w|_b}$ for every $w \in \Sigma^*$.

A *run* $r = q_0 \xrightarrow{a_1} q_1 \xrightarrow{a_2} \dots \xrightarrow{a_k} q_k$ of $\mathcal{A}$ is a sequence of transitions such that $(q_{i-1}, a_i, q_i) \in \Delta$ for each $i \in \{1, ..., k\}$. The *label* of this run is $a_1...a_k$. The *running weight* of this run is the product

$$\mathsf{rwt}(r) = \mathsf{in}(q_0) \cdot \prod_{1 \leq i \leq k} \mathsf{wt}(q_{i-1}, a_i, q_i) \cdot \mathsf{out}(q_k).$$

If $k = 0$ and $r = q_0$, then we define $\mathsf{rwt}(r) = \mathsf{in}(q_0) \cdot \mathsf{out}(q_0)$. The behaviour of a weighted finite automaton $\mathcal{A}$ is the series defined by

$$(\|\mathcal{A}\|, w) = \sum \{\mathsf{rwt}(r) : r \text{ is a run of } \mathcal{A} \text{ on } w\}.$$

A series $S \in \mathcal{K}\langle\langle\Sigma^*\rangle\rangle$ is called *WFA-recognizable* if there is a weighted finite automaton $\mathcal{A}$ such that $\|\mathcal{A}\| = S$.

Similar to the classical theory of formal languages, there are several characterizations of WFA-recognizable series. Here, we are mainly interested in weighted extensions of the well known Kleene theorem [76] and, second, the Büchi theorem [36]. Schützenberger [100] came up with a Kleene theorem for the class of weighted finite automata, known as Schützenberger theorem. It states that the set of WFA-recognizable series is precisely the set of *rational* series. The latter are defined over a finite set of monomials (i.e., series whose support is empty or a singleton) and finite application of sum, Cauchy product and finite Kleene star iteration. A weighted version of the Büchi theorem was recently given by Droste and Gastin [46, 49]. They present a weighted MSO logic and show that a fragment of this logic is equally expressive to the behaviours of weighted finite automata.

Besides characterizations of WFA-recognizable series, there has been much research on the supports of series. Next, we present a basic result concerning supports.

Theorem 2.8 (cf. [19]). *The support of each WFA-recognizable series over a positive semiring is recognizable by a finite automaton.*

The proof idea for this uses the two properties of being zero-sum free and zero-divisor free. Given a weighted finite automaton $\mathcal{A} = (Q, \Delta, \mathsf{in}, \mathsf{out}, \mathsf{wt})$, we construct a finite automaton recognizing $\mathsf{supp}(\|\mathcal{A}\|)$ by simply removing all transitions with weight 0 and defining the set of initial states to be the set of states q with $\mathsf{in}(q) \neq 0$ and the set of final states to be the set of states q with $\mathsf{out}(q) \neq 0$, respectively (cf. [19]). This result was recently supplemented by the following result for the class of commutative and zero-sum free semirings.

Theorem 2.9 ([75]). *The support of each WFA-recognizable series over a commutative and zero-sum free semiring is recognizable by a finite automaton.*

The proof of this is a bit more elaborate, but we will explain its details in Sect. 6.

On the other hand, it is well known that for general semirings, the supports of WFA-recognizable series are not necessarily recognizable by a finite automaton, as the following example (taken from [19]) shows.

Example 2.10. The support of the series recognized by the weighted finite automaton in Fig. 2.2 corresponds to the set $\{w \in \Sigma^* : |w|_a \neq |w|_b\}$, which is a context-free, but not a recognizable language.

However, for certain semirings we can decide whether the support of a WFA-recognizable series is empty. Notice that this problem corresponds to a weighted version of the classical emptiness problem.

Theorem 2.11 ([58]). *It is decidable whether the support of a WFA-recognizable series over a field is empty.*

The proof for the last theorem uses an alternative representation of WFA-recognizable series, called *linear representation*, and algebraic methods for which it is not clear whether and how they can be adapted to the timed setting. It is also known that for a given WFA-recognizable series S over a field and Σ, it is not decidable whether $\mathsf{supp}(S) = \Sigma^*$ (cf. [19]). Notice that this is the weighted version of the classical universality problem.

3 Weighted Timed Automata

Weighted timed automata have first been introduced in 2001 by Alur et al. [8] and Behrmann et al. [17] independently of each other. The authors extend the classical timed automaton model introduced by Alur and Dill [6] with a function $\mathsf{wt} : E \cup \mathcal{L} \to \mathbb{N}$ assigning a weight to each edge and each location (see Fig. 3.1, taken from [8]). The weight assigned to an edge corresponds to the cost for taking this edge, whereas the weight assigned to a location determines the cost for staying in that location *per time unit*. Given a run of the form $(l_0, \nu_0) \xrightarrow{\delta_1} \xrightarrow{e_1} \ldots \xrightarrow{\delta_k} \xrightarrow{e_k} (l_k, \nu_k)$, the weight of this run is defined to be the sum $\sum_{1 \leq i \leq k} \mathsf{wt}(l_{i-1}) \cdot \delta_i + \mathsf{wt}(e_i)$. Note that the weights assigned to the edges and locations of the timed automaton do not influence the behaviour of the timed automaton, but may be used to measure the *performance* of a run. In this spirit, the authors of both papers are interested in the *optimal cost reachability problem*, i.e., given a location l, to answer the question what is the optimal cost for reaching l.

The outstanding and new feature of weighted timed automata is the possibility of modelling *continuous* resource consumption that may arise while being in a location. Weighted timed automata have received much interest as they allow for applications in operations research, in particular *optimal* scheduling and planning [64, 97, 45]. Consequently, the model has been the subject of much research work in the real-time community. Besides further work on reachability problems under some optimization aspect [16, 104, 22], there are several approaches to extend temporal logics like CTL and LTL with cost constraints on modalities [32, 21, 23, 34, 26]. Also, there are efforts [4, 21, 23] to generalize the notion of time-optimal games [15], which is of practical interest e.g. for solving optimal control problems, to the weighted timed setting. Another interesting direction are extensions of the model described above to so-called *multi-priced timed automata* [80, 24, 65], where multiple cost variables per transition (and location, respectively) are allowed. This model allows for practically relevant problems, particularly in the framework of embedded systems, where often more than one resource must

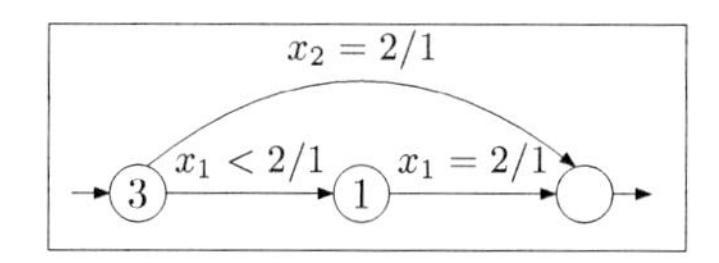

Figure 3.1: A weighted timed automaton as defined by Alur et al. [8]

be restricted.

In this section, we present a more general definition of a weighted timed automaton which is along the lines of the definition of a weighted finite automaton. By building a bridge to the theory of weighted finite automata, we hope to get a new, mainly theoretically-influenced perspective on weighted timed automata. This may be helpful in obtaining a deeper understanding about the model itself. The main difference to weighted finite automata is that not only the edges of the underlying timed automaton are assigned weights from a semiring, but also the locations are assigned a weight function, which is used to compute the weight that arises while letting time elapse in this location. Note that, compared to the model described above, in our model these weight functions may be non-linear. Due to the weight functions assigned to locations, our model of a weighted timed automaton is far more complicated than that of a weighted finite automaton. In particular, there may be an infinite set of weights occurring in the runs of a weighted timed automaton. Next, we present the formal definition of a weighted timed automaton.

Let $\mathcal{K}$ be a semiring. We use $\mathcal{F}$ to denote a family of functions from $\mathbb{R}_{\geq 0}$ to K which will be used to model the weights that arise while being in a location. In this thesis, we are mainly interested in two kinds of such families, namely the family of *step functions* and the family of *linear functions*. A function $f : \mathbb{R}_{\geq 0} \to K$ is a *step function* if it is of the form $f(\delta) = \sum_{1 \leq i \leq n} \alpha_i \cdot \chi_{A_i}(\delta)$ for every $\delta \in \mathbb{R}_{\geq 0}$, where $n \in \mathbb{N}$, $\alpha_i \in K$, A_i are intervals over $\mathbb{R}_{\geq 0}$ with borders in $\mathbb{N}$ such that $A_j \cap A_k = \emptyset$ for $j \neq k$ and $\bigcup_{1 \leq i \leq n} A_i = \mathbb{R}_{\geq 0}$, and χ_{A_i} is a characteristic function of A_i, i.e., we have $\chi_{A_i}(\delta) = 1$ if $\delta \in A_i$ and 0 otherwise, for every $i \in \{1, ..., n\}$. A step function has the important property of having a finite image. *Constant functions* of the form $f(\delta) = k$ for some $k \in \mathcal{K}$ and each $\delta \in \mathbb{R}_{\geq 0}$, are a special case of step functions. If $\mathcal{K}$ is such that $K \supseteq \mathbb{R}_{\geq 0}$, we say that a function $f : \mathbb{R}_{\geq 0} \to K$ is *linear* if it is of the form $f(\delta) = k \cdot \delta$ for some $k \in K \cap \mathbb{R}$ and every $\delta \in \mathbb{R}_{\geq 0}$ (where $\cdot$ is the usual multiplication operation). In this thesis, we often require $\mathcal{F}$ to contain a special function $f : \mathbb{R}_{\geq 0} \to K$ satisfying $f(\delta) = 1$ for each $\delta \in \mathbb{R}_{\geq 0}$. We use $\mathbb{1}$ to denote such a function. Given two functions $f_1, f_2 \in \mathcal{F}$, we define the *pointwise product* $f_1 \odot f_2$ by $(f_1 \odot f_2)(\delta) = f_1(\delta) \cdot f_2(\delta)$ for each $\delta \in \mathbb{R}_{\geq 0}$.

A *weighted timed automaton* over $\mathcal{K}$, Σ and $\mathcal{F}$ is a tuple $\mathcal{A} = (\mathcal{L}, \mathcal{C}, E, \mathsf{in}, \mathsf{out}, \mathsf{ewt}, \mathsf{lwt})$, where

- $\mathcal{L}$ is a finite set of locations,
- $\mathcal{C}$ is a finite set of clock variables,
- $E \subseteq \mathcal{L} \times \Sigma \times \Phi(\mathcal{C}) \times 2^{\mathcal{C}} \times \mathcal{L}$ is a finite set of edges,
- $\mathsf{in} : \mathcal{L} \to K$ is a weight function for entering a location,
- $\mathsf{out} : \mathcal{L} \to K$ is a weight function for leaving a location,

- $\mathsf{ewt} : E \to K$ is a weight function for taking an edge,
- $\mathsf{lwt} : \mathcal{L} \to \mathcal{F}$ is a weight function for letting time elapse in a location.

For representing weighted timed automata graphically, we use the same conventions as for weighted finite automata (see Sect. 2.3). Additionally, we label the locations with their weight functions, i.e., if $\mathsf{lwt}(l) = f$, then we label l with f.

A weighted timed automaton maps each timed word $w \in T\Sigma^*$ to a weight in K as follows. Given a run $r = (l_0, \nu_0) \xrightarrow{\delta_1} \xrightarrow{e_1} \dots \xrightarrow{\delta_k} \xrightarrow{e_k} (l_k, \nu_k)$ of $\mathcal{A}$ on w, we define the *running weight* $\mathsf{rwt}(r)$ of r to be the product

$$\mathsf{rwt}(r) = \mathsf{in}(l_0) \cdot \left(\prod_{1 \le i \le k} \mathsf{lwt}(l_{i-1})(\delta_i) \cdot \mathsf{ewt}(e_i) \right) \cdot \mathsf{out}(l_k).$$

Then the behaviour $\|\mathcal{A}\| : T\Sigma^* \to K$ of $\mathcal{A}$ is given by

$$(\|\mathcal{A}\|, w) = \sum \{\mathsf{rwt}(r) : r \text{ is a run of } \mathcal{A} \text{ on } w\}.$$

A function $\mathcal{T} : T\Sigma^* \to K$ is called *timed series*. A timed series $\mathcal{T}$ is called *$\mathcal{F}$-recognizable* over $\mathcal{K}$ and Σ if there is a weighted timed automaton $\mathcal{A}$ over $\mathcal{K}$, Σ and $\mathcal{F}$ such that $\|\mathcal{A}\| = \mathcal{T}$. If $\mathcal{K}$, Σ or $\mathcal{F}$ are clear from the context, we may omit them. We use $\mathcal{K}^{\mathcal{F}-rec}\langle\langle T\Sigma^* \rangle\rangle$ to denote the set of all $\mathcal{F}$-recognizable timed series over $\mathcal{K}$ and Σ.

Example 3.1. In Fig. 3.2, we show a weighted timed automaton over the semiring of the natural numbers with addition and multiplication and the family of constant functions. In fact, we only use the constant function $\mathbb{1}$, which maps each time delay to 1. Also the edges are assigned the weight 1. From this we can conclude that the running weight of each run from the leftmost to the rightmost location equals 1. Moreover, for each timed word w, there are exactly as many such runs on w as there are positions in w such that the last a happened strictly less than 2 time units ago. The behaviour of the weighted timed automaton with respect to w is computed by summing up the running weights of the runs on w. Thus, this weighted timed automaton assigns to each timed word the number of positions such that the last a happened less than 2 time units ago. In general, one may use the semiring of the natural numbers to *count how often* a certain property holds.

Example 3.2. Let $\mathcal{K}$ be the max-plus-semiring and $\mathcal{F}$ be the family of linear and constant functions. Consider the weighted timed automaton $\mathcal{A}$ over $\mathcal{K}$ and $\mathcal{F}$ shown in Fig. 3.3. Let w be a timed word. Then for each $i \in \mathsf{dom}(w)$, there is a run r on w such that $\mathsf{rwt}(r) = t_i - t_{i-1}$ (where $t_0 = 0$). Moreover, there are no other runs. Hence, we have $(\|\mathcal{A}\|, w) = \max\{t_i - t_{i-1} : i \in \mathsf{dom}(w)\}$.

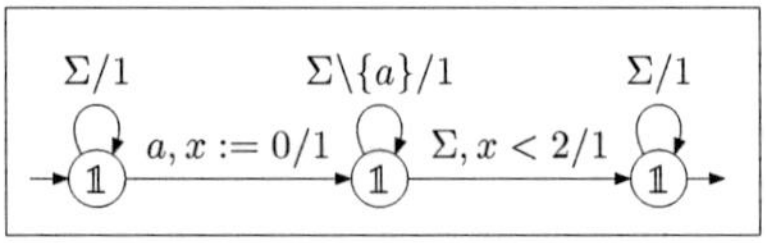

Figure 3.2: Weighted timed automaton for Example 3.1

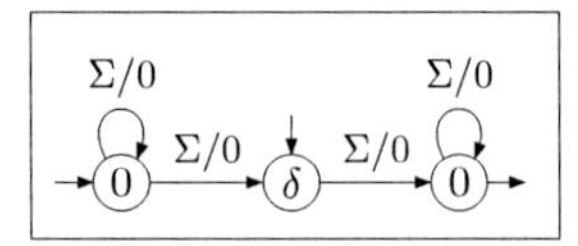

Figure 3.3: Weighted timed automaton for Example 3.2

A timed series $\mathcal{T}$ is called *strictly monotonic* if for each timed word $w \in T\Sigma^* \backslash T_s\Sigma^*$ we have $(\mathcal{T}, w) = 0$. We define the (potentially infinite) set $\mathsf{wgt}(\mathcal{A})$ of weights that may occur in the runs of $\mathcal{A}$ as $\mathsf{wgt}(\mathcal{A}) = \mathsf{wgt}_E(\mathcal{A}) \cup \mathsf{wgt}_{\mathcal{F}}(\mathcal{A})$, where $\mathsf{wgt}_E(\mathcal{A}) = \{\mathsf{ewt}(e) : e \in E\}$ and $\mathsf{wgt}_{\mathcal{F}}(\mathcal{A}) = \{\mathsf{lwt}(l)(\delta) : l \in \mathcal{L}, \delta \in \mathbb{R}_{\geq 0}\}$. Observe that a weighted timed automaton $\mathcal{A}$ over $\mathcal{K}$, Σ and $\mathcal{F}$ can also be regarded as a weighted timed automaton over the subsemiring $\mathcal{K}_{\mathsf{wgt}(\mathcal{A})}$, Σ and $\mathcal{F}$. Given a weighted timed automaton of $\mathcal{A} = (\mathcal{L}, \mathcal{C}, E, \mathsf{in}, \mathsf{out}, \mathsf{ewt}, \mathsf{lwt})$ over $\mathcal{K}$, Σ and $\mathcal{F}$, the *underlying* timed automaton of $\mathcal{A}$ is the timed automaton $(\mathcal{L}, \mathcal{L}_0, \mathcal{L}_f, \mathcal{C}, E)$ over Σ where $\mathcal{L}_0 = \{l \in \mathcal{L} : \mathsf{in}(l) \neq 0\}$ and $\mathcal{L}_f = \{l \in \mathcal{L} : \mathsf{out}(l) \neq 0\}$. We say that a weighted timed automaton is *unambiguous* if its underlying timed automaton is unambiguous.

3.1 Relation To Other Automata Models

Here we show that our model of a weighted timed automaton over a semiring and a family of functions subsumes a number of more particular concepts of timed automata. It is well known that weighted timed automata can be seen as an extension of timed automata towards linear hybrid systems [70, 5]. However, our definition of a weighted timed automaton also includes various other classes of automata. In particular, by choosing $\mathcal{K}$ and $\mathcal{F}$ in a suitable way, we obtain timed automata and weighted finite automata.

Timed Automata Let $\mathcal{A} = (\mathcal{L}, \mathcal{L}_0, \mathcal{L}_f, \mathcal{C}, E)$ be a timed automaton over Σ. We let $\mathcal{K}$ be the Boolean semiring and $\mathcal{F} = \{\mathbb{1}\}$. Define the weight function $\mathsf{in} : \mathcal{L} \to K$ by $\mathsf{in}(l) = 1$ if $l \in \mathcal{L}_0$ and $\mathsf{in}(l) = 0$ otherwise. Similarly, define $\mathsf{out} : \mathcal{L} \to K$ by $\mathsf{out}(l) = 1$ if $l \in \mathcal{L}_f$ and $\mathsf{out}(l) = 0$ otherwise. Further, we let $\mathsf{ewt}(e) = 1$ for each $e \in E$ and $\mathsf{lwt}(l) = \mathbb{1}$ for each $l \in \mathcal{L}$. Let $\mathcal{A}' = (\mathcal{L}, \mathcal{C}, E, \mathsf{in}, \mathsf{out}, \mathsf{ewt}, \mathsf{lwt})$ be a weighted timed automaton over

$\mathcal{K}$, Σ and $\mathcal{F}$. One can easily see that $\|\mathcal{A}'\| = 1_{L(\mathcal{A})}$.

Other Weighted Timed Automata Models In the original work on weighted timed automata by Alur et al. [8] and Behrmann et al. [17], timed automata are extended with a weight function assigning a natural number to each edge and each location. The weights that arise by letting time elapse while being in a location are computed by multiplying the weight of the location with the time value. The weights of the transitions occuring in a run are added. Furthermore, the run with *minimal* weight is of interest. This can be modelled using our definition of a weighted timed automaton over the min-plus-semiring and the family of linear functions.

Recently, the weighted timed automaton model has been generalized by allowing more than one weight variable. Larsen and Rasmussen introduced dual-priced timed automata [80]. A dual-priced timed automaton can be modelled using a weighted timed automaton over the direct product of the min-plus-semiring, i.e., $((\mathbb{R}_{\geq 0} \cup \{\infty\}) \times (\mathbb{R}_{\geq 0} \cup \{\infty\}), \min, +, (\infty, \infty), (0, 0))$, where min and + are defined in a suitable way, e.g. coordinate-wise. This again constitutes a semiring and thus falls into our framework. Similarly, we can define models of multi-priced timed automata [24].

Timed Automata with Stopwatch Observers A stopwatch is a clock variable that can be stopped and turned on again [70]. In other words, the rate of change of the stopwatch variable is either $0 \in \mathbb{N}$ or $1 \in \mathbb{N}$. A timed automaton augmented with a stopwatch variable that can neither be tested in a clock constraint nor be reset is called a timed automaton with a stopwatch observer. To model this using our definition of a weighted timed automaton, we let $\mathcal{F}$ contain $\mathbb{1}$ and the linear function f of the form $f(\delta) = \delta$. This implies that $\mathcal{K}$ must be such that $K \supseteq \mathbb{R}_{\geq 0}$, but notice that we do not put further restrictions on $\mathcal{K}$ here, but describe the modelling for the general case. We further let $\mathsf{lwt}(l) = \mathbb{1}$ if l is a location where the stopwatch is stopped, and $\mathsf{lwt}(l) = f$ if in l the stopwatch is turned on. The edges do not cost anything, so we put $\mathsf{ewt}(e) = 1$ for any $e \in E$. The weight function for entering a location is defined by $\mathsf{in}(l) = 1$ if $l \in \mathcal{L}_0$ and $\mathsf{in}(l) = 0$ otherwise, and similarly we proceed with the weight function for leaving a location.

Weighted Finite Automata A weighted finite automaton over a semiring $\mathcal{K}$ and Σ can easily be modelled by a weighted timed automaton over $\mathcal{K}$, Σ and the family of functions containing $\mathbb{1}$. By assigning $\mathbb{1}$ to each location of a weighted timed automaton, no weight arises while being in a location. In this way, we obtain a weighted timed automaton model that allows for weights at the edges. If we additionally ignore all timing information, the resulting model corresponds to a classical weighted finite automaton.

3.2 Closure Properties of Recognizable Timed Series

Here, we investigate closure properties of the class of $\mathcal{F}$-recognizable timed series under some standard operations which will be needed later in this thesis. For this, we fix a semiring $\mathcal{K}$, an alphabet Σ and a family $\mathcal{F}$ of functions from $\mathbb{R}_{\geq 0}$ to K. The operations of sum, Hadamard product and scalar products are lifted from series to timed series in the obvious manner, i.e., given timed series $\mathcal{T}, \mathcal{T}_1, \mathcal{T}_2 : T\Sigma^* \to K$ and $k \in K$, we define

$$\begin{aligned}(\mathcal{T}_1 + \mathcal{T}_2, w) &= (\mathcal{T}_1, w) + (\mathcal{T}_2, w),\\ (\mathcal{T}_1 \odot \mathcal{T}_2, w) &= (\mathcal{T}_1, w) \cdot (\mathcal{T}_2, w),\\ (\mathcal{T} \cdot k, w) &= (\mathcal{T}, w) \cdot k,\\ (k \cdot \mathcal{T}, w) &= k \cdot (\mathcal{T}, w).\end{aligned}$$

Given a mapping $\pi : \Sigma \to \Gamma$ and a timed series $\mathcal{T} : T\Sigma^* \to K$, we define the renaming $\bar{\pi}(\mathcal{T}) : T\Gamma^* \to K$ by $(\bar{\pi}(\mathcal{T}), u) = \sum_{\pi(w)=u} (\mathcal{T}, w)$ for all $u \in T\Gamma^*$, and for a timed series $\mathcal{T} : T\Gamma^* \to K$ we define the inverse renaming $\bar{\pi}^{-1}(\mathcal{T}) : T\Sigma^* \to K$ by $(\bar{\pi}^{-1}(\mathcal{T}), w) = (\mathcal{T}, \pi(w))$ for each $w \in T\Sigma^*$. The proof for closure of the class of $\mathcal{F}$-recognizable timed series under sum can be done as the proof for closure of the class of classical recognizable languages under union, namely by taking a disjoint union of two weighted timed automata.

Lemma 3.3. *The class of $\mathcal{F}$-recognizable timed series is closed under sum.*

PROOF. Let $\mathcal{T}_1, \mathcal{T}_2 : T\Sigma^* \to K$ be $\mathcal{F}$-recognizable timed series. Then there are weighted timed automata $\mathcal{A}_1 = (\mathcal{L}_1, \mathcal{C}_1, E_1, \mathsf{in}_1, \mathsf{out}_1, \mathsf{ewt}_1, \mathsf{lwt}_1)$ and $\mathcal{A}_2 = (\mathcal{L}_2, \mathcal{C}_2, E_2, \mathsf{in}_2, \mathsf{out}_2, \mathsf{ewt}_2, \mathsf{lwt}_2)$ over $\mathcal{K}$, Σ and $\mathcal{F}$ such that $\|\mathcal{A}_1\| = \mathcal{T}_1$ and $\|\mathcal{A}_2\| = \mathcal{T}_2$, respectively. Without loss of generality, we may assume $\mathcal{L}_1 \cap \mathcal{L}_2 = \emptyset$ and $\mathcal{C}_1 \cap \mathcal{C}_2 = \emptyset$. Define $\mathcal{A} = (\mathcal{L}, \mathcal{C}, E, \mathsf{in}, \mathsf{out}, \mathsf{ewt}, \mathsf{lwt})$, where

- $\mathcal{L} = \mathcal{L}_1 \cup \mathcal{L}_2$, $\mathcal{C} = \mathcal{C}_1 \cup \mathcal{C}_2$, $E = E_1 \cup E_2$,
- $\mathsf{in} = \mathsf{in}_1 \cup \mathsf{in}_2$, $\mathsf{out} = \mathsf{out}_1 \cup \mathsf{out}_2$, $\mathsf{ewt} = \mathsf{ewt}_1 \cup \mathsf{ewt}_2$, $\mathsf{lwt} = \mathsf{lwt}_1 \cup \mathsf{lwt}_2$.

Then $\|\mathcal{A}\| = \|\mathcal{A}_1\| + \|\mathcal{A}_2\|$ follows from the fact that for each $w \in T\Sigma^*$, every run of $\mathcal{A}$ on w is a run of $\mathcal{A}_1$ or a run of $\mathcal{A}_2$ on w with the same running weight, and vice versa. ■

Unfortunately, the proof for closure of the class of $\mathcal{F}$-recognizable timed series under the Hadamard product cannot be adopted so easily from the corresponding classical case, i.e., closure of recognizable languages under intersection. In fact, in general the class of $\mathcal{F}$-recognizable timed series is not closed under the Hadamard product due to two reasons. First, as in the untimed setting [49], we must ensure that the weights occuring in runs of the two weighted timed automata commute element-wise. This can be solved by assuming $\mathcal{K}$ to be commutative. Second, we have to restrict the location weight functions used in the weighted timed automata in order to guarantee that the pointwise product of each pair of functions is in $\mathcal{F}$. This is illustrated in the next example.

Example 3.4. Let $\mathcal{K}$ be commutative and such that $K \supseteq \mathbb{R}_{\geq 0}$, let $\Sigma = \{a\}$ and let $\mathcal{F}$ be the family of linear functions. In this example, we use $\bullet$ to denote the multiplication of $\mathcal{K}$ and $\cdot$ to denote the usual multiplication. Let $\mathcal{A} = (\mathcal{L}, \mathcal{C}, E, \mathsf{in}, \mathsf{out}, \mathsf{ewt}, \mathsf{lwt})$ be a weighted timed automaton over $\mathcal{K}$, Σ and $\mathcal{F}$, where

- $\mathcal{L} = \{l_1, l_2\}$,
- $\mathcal{C} = \emptyset$,
- $E = \{(l_1, a, \mathsf{true}, \emptyset, l_2)\}$ and $\mathsf{ewt}(l_1, a, \mathsf{true}, \emptyset, l_2) = 1$,
- $\mathsf{in}(l_1) = 1$ and $\mathsf{in}(l_2) = 0$,
- $\mathsf{out}(l_1) = 0$ and $\mathsf{out}(l_2) = 1$,
- $\mathsf{lwt}(l_1)(\delta) = 2 \cdot \delta$ for each $\delta \in \mathbb{R}_{\geq 0}$, $\mathsf{lwt}(l_2) \in \mathcal{F}$ is arbitrary.

We further let $\mathcal{A}'$ be a copy of $\mathcal{A}$, except for lwt' which is defined by $\mathsf{lwt}'(l_1')(\delta) = 3 \cdot \delta$ for each $\delta \in \mathbb{R}_{\geq 0}$.

Let $w \in T\Sigma^*$. If $w \neq (a, t)$ for some $t \in \mathbb{R}_{\geq 0}$, then $(\|\mathcal{A}\|, w) = (\|\mathcal{A}'\|, w) = 0$ and thus $(\|\mathcal{A}\| \odot \|\mathcal{A}'\|, w) = 0$. So let $w = (a, t)$ for some $t \in \mathbb{R}_{\geq 0}$. Then we have $(\|\mathcal{A}\| \odot \|\mathcal{A}'\|, w) = 2 \cdot t \bullet 3 \cdot t$.

Now, if $\mathcal{K}$ is the min-plus-semiring, we obtain $(\|\mathcal{A}\| \odot \|\mathcal{A}'\|, w) = 2 \cdot t + 3 \cdot t = 5 \cdot t$. Clearly, this timed series is $\mathcal{F}$-recognizable. The idea is to use the usual product automaton construction and define the weight functions using the pointwise product of the corresponding weight functions of $\mathcal{A}$ and $\mathcal{A}'$. This can be done since the pointwise product of each pair of linear functions is a linear function and thus in $\mathcal{F}$.

However, if $\mathcal{K}$ is the semiring of the real numbers with addition and multiplication, we have $(\|\mathcal{A}\| \odot \|\mathcal{A}'\|, w) = 2 \cdot t \cdot 3 \cdot t = 6 \cdot t^2$. It can be easily seen that there is no weighted timed automaton over the family of *linear* functions recognizing $\|\mathcal{A}\| \odot \|\mathcal{A}'\|$.

For this reason, we define the notion of *non-interfering* timed series. Let $\mathcal{A} = (\mathcal{L}, \mathcal{C}, E, \mathsf{in}, \mathsf{out}, \mathsf{ewt}, \mathsf{lwt})$ and $\mathcal{A}' = (\mathcal{L}', \mathcal{C}', E', \mathsf{in}', \mathsf{out}', \mathsf{ewt}', \mathsf{lwt}')$ be two weighted timed automata over $\mathcal{K}$, Σ and $\mathcal{F}$. Define $\mathcal{L}_f = \{(l, l') \in \mathcal{L} \times \mathcal{L}' : \mathsf{out}(l) \neq 0, \mathsf{out}'(l') \neq 0\}$. We say that $\mathcal{A}$ and $\mathcal{A}$ are *non-interfering* if for all pairs $(l, l') \in \mathcal{L} \times \mathcal{L}'$, whenever there is a run from (l, l') into $\mathcal{L}_f$, then $\mathsf{lwt}(l) \odot \mathsf{lwt}'(l') \in \mathcal{F}$. This guarantees that the product automaton of $\mathcal{A}$ and $\mathcal{A}'$ is a weighted timed automaton over $\mathcal{F}$. If $\mathcal{F}$ is closed under the pointwise product, all pairs of weighted timed automata are non-interfering. However, the condition is also satisfied, if $\mathsf{lwt}(l) = \mathbb{1}$ or $\mathsf{lwt}'(l') = \mathbb{1}$ for each pair $(l, l') \in \mathcal{L} \times \mathcal{L}'$

from which there is a run into $\mathcal{L}_f$. Also notice that the premise of the condition - testing for reachability of locations - is decidable [6]. Two timed series $\mathcal{T}, \mathcal{T}' : T\Sigma^* \to K$ are *non-interfering* over $\mathcal{K}$, Σ and $\mathcal{F}$ if there are two non-interfering weighted timed automata $\mathcal{A}$ and $\mathcal{A}'$ over $\mathcal{K}$, Σ and $\mathcal{F}$ such that $\|\mathcal{A}\| = \mathcal{T}$ and $\|\mathcal{A}'\| = \mathcal{T}'$.

Lemma 3.5. *Let $\mathcal{K}$ be commutative. If $\mathcal{T}_1, \mathcal{T}_2 : T\Sigma^* \to K$ are non-interfering timed series over $\mathcal{K}$, Σ and $\mathcal{F}$, then $\mathcal{T}_1 \odot \mathcal{T}_2$ is $\mathcal{F}$-recognizable over $\mathcal{K}$ and Σ.*

Proof. Let $\mathcal{K}$ be commutative and $\mathcal{T}_1, \mathcal{T}_2 : T\Sigma^* \to K$ be non-interfering timed series over $\mathcal{K}$, Σ and $\mathcal{F}$. Then there exist two non-interfering weighted timed automata $\mathcal{A}_i = (\mathcal{L}_i, \mathcal{C}_i, E_i, \mathsf{in}_i, \mathsf{out}_i, \mathsf{ewt}_i, \mathsf{lwt}_i)$ over $\mathcal{K}$, Σ and $\mathcal{F}$ $(i = 1, 2)$ such that $\|\mathcal{A}_1\| = \mathcal{T}_1$ and $\|\mathcal{A}_2\| = \mathcal{T}_2$. We may assume that $\mathcal{L}_1 \cap \mathcal{L}_2 = \emptyset$ and $\mathcal{C}_1 \cap \mathcal{C}_2 = \emptyset$. Define $\mathcal{L}_{\text{bad}} = \{(l_1, l_2) \in \mathcal{L}_1 \times \mathcal{L}_2 : \mathsf{lwt}_1(l_1) \odot \mathsf{lwt}_2(l_2) \notin \mathcal{F}\}$. We let $\mathcal{A} = (\mathcal{L}, \mathcal{C}, E, \mathsf{in}, \mathsf{out}, \mathsf{ewt}, \mathsf{lwt})$ be the weighted timed automaton over $\mathcal{K}$, Σ and $\mathcal{F}$ defined by

- $\mathcal{L} = (\mathcal{L}_1 \times \mathcal{L}_2) \backslash \mathcal{L}_{\text{bad}}$,
- $\mathcal{C} = \mathcal{C}_1 \cup \mathcal{C}_2$,
- $E = \{((l_1, l_2), a, \phi_1 \wedge \phi_2, \lambda_1 \cup \lambda_2, (l_1', l_2')) \in E : (l_1, a, \phi_1, \lambda_1, l_1') \in E_1, (l_2, a, \phi_2, \lambda_2, l_2') \in E_2, (l_1, l_2), (l_1', l_2') \in \mathcal{L}\}$,
- $\mathsf{in}((l_1, l_2)) = \mathsf{in}_1(l_1) \cdot \mathsf{in}_2(l_2)$ for each $(l_1, l_2) \in \mathcal{L}$,
- $\mathsf{out}((l_1, l_2)) = \mathsf{out}_1(l_1) \cdot \mathsf{out}_2(l_2)$ for each $(l_1, l_2) \in \mathcal{L}$,
- $\mathsf{ewt}((l_1, l_2), a, \phi_1 \wedge \phi_2, \lambda_1 \cup \lambda_2, (l_1', l_2')) = \mathsf{ewt}_1(l_1, a, \phi_1, \lambda_1, l_1') \cdot \mathsf{ewt}_2(l_2, a, \phi_2, \lambda_2, l_2')$ for each $((l_1, l_2), a, \phi_1 \wedge \phi_2, \lambda_1 \cup \lambda_2, (l_1', l_2')) \in E$,
- $\mathsf{lwt}((l_1, l_2)) = \mathsf{lwt}_1(l_1) \odot \mathsf{lwt}_2(l_2)$ for every $(l_1, l_2) \in \mathcal{L}$.

Intuitively, $\mathcal{A}$ is the classical product automaton, but we remove all "bad" pairs of locations whose pointwise product of location weight functions is not in $\mathcal{F}$. As a consequence, we obtain $\mathsf{lwt}((l_1, l_2)) \in \mathcal{F}$ for every $(l_1, l_2) \in \mathcal{L}$. The removing of "bad" pairs of locations can be done since by assumption from every such pair there is no run into $\mathcal{L}_f$ anyway. Subsequently, we show that $\|\mathcal{A}\| = \|\mathcal{A}_1\| \odot \|\mathcal{A}_2\|$. We start by proving that there is a weight-preserving bijective correspondence between the set of runs of $\mathcal{A}$ and the set of pairs of runs of $\mathcal{A}_1$ and $\mathcal{A}_2$. Let $w \in T\Sigma^*$. Suppose there is a run $r = ((l_0, l_0'), \nu_0) \xrightarrow{\delta_1} \xrightarrow{e_1} \dots \xrightarrow{\delta_{|w|}} \xrightarrow{e_{|w|}} ((l_{|w|}, l_{|w|}'), \nu_{|w|})$ of $\mathcal{A}$ on w. The construction of $\mathcal{A}$ implies that there are runs $r_1 = (l_0, \nu_{0|\mathcal{C}_1}) \xrightarrow{\delta_1} \xrightarrow{e_1^1} \dots \xrightarrow{\delta_{|w|}} \xrightarrow{e_{|w|}^1} (l_{|w|}, \nu_{|w||\mathcal{C}_1})$ of $\mathcal{A}_1$ and $r_2 = (l_0', \nu_{0|\mathcal{C}_2}) \xrightarrow{\delta_1} \xrightarrow{e_1^2} \dots \xrightarrow{\delta_{|w|}} \xrightarrow{e_{|w|}^2} (l_{|w|}', \nu_{|w||\mathcal{C}_2})$ of $\mathcal{A}_2$, respectively, on w. Using the definition of the weight functions in $\mathcal{A}$ and commutativity of $\mathcal{K}$, we obtain

$$
\begin{aligned}
& \mathsf{rwt}(r) \\
= \; & \mathsf{in}((l_0, l'_0)) \cdot \left(\prod_{1 \le i \le |w|} \mathsf{lwt}((l_{i-1}, l'_{i-1}))(\delta_i) \cdot \mathsf{ewt}(e_i) \right) \cdot \mathsf{out}((l_{|w|}, l'_{|w|})) \\
= \; & \mathsf{in}_1(l_0) \cdot \mathsf{in}_2(l'_0) \cdot \left(\prod_{1 \le i \le |w|} \mathsf{lwt}_1(l_{i-1})(\delta_i) \cdot \mathsf{lwt}_2(l'_{i-1})(\delta_i) \cdot \mathsf{ewt}_1(e_i^1) \cdot \mathsf{ewt}_2(e_i^2) \right) \\
& \cdot \mathsf{out}_1(l_{|w|}) \cdot \mathsf{out}_2(l'_{|w|}) \\
= \; & \mathsf{in}_1(l_0) \cdot \left(\prod_{1 \le i \le |w|} \mathsf{lwt}_1(l_{i-1})(\delta_i) \cdot \mathsf{ewt}_1(e_i^1) \right) \cdot \mathsf{out}_1(l_{|w|}) \\
& \cdot \mathsf{in}_2(l'_0) \cdot \left(\prod_{1 \le i \le |w|} \mathsf{lwt}_2(l'_{i-1})(\delta_i) \cdot \mathsf{ewt}_2(e_i^2) \right) \cdot \mathsf{out}_2(l'_{|w|}) \\
= \; & \mathsf{rwt}(r_1) \cdot \mathsf{rwt}(r_2).
\end{aligned}
$$

For the other direction, assume that r_1 and r_2 as above are runs of $\mathcal{A}_1$ and $\mathcal{A}_2$, respectively, on w. We distinguish between two cases. First, suppose there is some $i \in \mathsf{dom}(w)$ such that $\mathsf{lwt}_1(l_i) \odot \mathsf{lwt}_2(l'_i) \notin \mathcal{F}$. By the assumption that $\mathcal{A}_1$ and $\mathcal{A}_2$ are non-interfering over $\mathcal{K}$, Σ and $\mathcal{F}$, we know that from (l_i, l'_i) there is no run into $\mathcal{L}_f$. But this means that for all $j \in \{i, ..., |w|\}$, we have $\mathsf{out}_1(l_j) = 0$ or $\mathsf{out}_2(l'_j) = 0$. Hence, either $\mathsf{rwt}(r_1) = 0$ or $\mathsf{rwt}(r_2) = 0$, and thus $\mathsf{rwt}(r_1) \cdot \mathsf{rwt}(r_2) = 0$. Since $(l_i, l'_i) \in \mathcal{L}_{\text{bad}}$, the composition of r_1 and r_2 is not a run of $\mathcal{A}$ on w. Second, suppose that $\mathsf{lwt}_1(l_i) \odot \mathsf{lwt}_2(l'_i) \in \mathcal{F}$ for each $i \in \mathsf{dom}(w)$. Then $(l_i, l'_i) \in \mathcal{L}$ for each $i \in \mathsf{dom}(w)$ and we can compose r_1 and r_2 to a run r of $\mathcal{A}$ on w. Using the same lines of argumentation as above, we can show $\mathsf{rwt}(r) = \mathsf{rwt}(r_1) \cdot \mathsf{rwt}(r_2)$. We use these two constructions in step $\star$ to finally show that for each $w \in T\Sigma^*$ we have

$$
\begin{aligned}
& (\|\mathcal{A}\|, w) \\
= \; & \sum \{\mathsf{rwt}(r) : r \text{ is a run of } \mathcal{A} \text{ on } w\} \\
\overset{\star}{=} \; & \sum \{\mathsf{rwt}(r_1) \cdot \mathsf{rwt}(r_2) : r_1 \text{ is a run of } \mathcal{A}_1 \text{ on } w, r_2 \text{ is a run of } \mathcal{A}_2 \text{ on } w\} \\
= \; & \sum \{\mathsf{rwt}(r_1) : r_1 \text{ is a run of } \mathcal{A}_1 \text{ on } w\} \cdot \sum \{\mathsf{rwt}(r_2) : r_1 \text{ is a run of } \mathcal{A}_2 \text{ on } w\} \\
= \; & (\|\mathcal{A}_1\|, w) \cdot (\|\mathcal{A}_2\|, w) \\
= \; & (\|\mathcal{A}_1\| \odot \|\mathcal{A}_2\|, w).
\end{aligned}
$$

■

Lemma 3.6. *The class of $\mathcal{F}$-recognizable timed series is closed under scalar products.*

PROOF. Let $\mathcal{T} : T\Sigma^* \to K$ be an $\mathcal{F}$-recognizable timed series and $k \in K$. Then there is a weighted timed automaton $\mathcal{A} = (\mathcal{L}, \mathcal{C}, E, \mathsf{in}, \mathsf{out}, \mathsf{ewt}, \mathsf{lwt})$ over $\mathcal{K}$, Σ and $\mathcal{F}$ such that $\|\mathcal{A}\| = \mathcal{T}$. Define $\mathcal{A}' = (\mathcal{L}, \mathcal{C}, E, \mathsf{in}', \mathsf{out}, \mathsf{ewt}, \mathsf{lwt})$, where $\mathsf{in}'(l) = k \cdot \mathsf{in}(l)$ for each $l \in \mathcal{L}$. Then we obtain for every $w \in T\Sigma^*$

$$\begin{aligned}
(\|\mathcal{A}'\|, w) &= \sum\{\mathsf{rwt}(r) : r \text{ is a run of } \mathcal{A}' \text{ on } w\} \\
&= \sum\{k \cdot \mathsf{rwt}(r) : r \text{ is a run of } \mathcal{A} \text{ on } w\} \\
&= k \cdot \sum\{\mathsf{rwt}(r) : r \text{ is a run of } \mathcal{A} \text{ on } w\} \\
&= k \cdot (\|\mathcal{A}\|, w)
\end{aligned}$$

what was to be demonstrated. The proof for $\mathcal{F}$-recognizability of $\mathcal{T} \cdot k$ can be done analogously by defining a new function out' by $\mathsf{out}'(l) = \mathsf{out}(l) \cdot k$ for each $l \in \mathcal{L}$. ■

Lemma 3.7. *The class of $\mathcal{F}$-recognizable timed series is closed under renamings.*

PROOF. Let $\mathcal{T} : T\Sigma^* \to K$ be an $\mathcal{F}$-recognizable timed series, and $\pi : \Sigma \to \Gamma$ be a mapping. Then, there is a weighted timed automaton $\mathcal{A} = (\mathcal{L}, \mathcal{C}, E, \mathsf{in}, \mathsf{out}, \mathsf{ewt}, \mathsf{lwt})$ over $\mathcal{K}$, Σ and $\mathcal{F}$ such that $\|\mathcal{A}\| = \mathcal{T}$. Define $E' = \{(l, \pi(a), \phi, \lambda, l') : (l, a, \phi, \lambda, l') \in E\}$. Now, define $\mathsf{ewt}' : E' \to K$ by

$$\mathsf{ewt}'(l, b, \phi, \lambda, l') = \sum_{\substack{(l,a,\phi,\lambda,l') \in E \\ \pi(a) = b}} \mathsf{ewt}(l, a, \phi, \lambda, l')$$

and put $\mathcal{A}' = (\mathcal{L}, \mathcal{C}, E', \mathsf{in}, \mathsf{out}, \mathsf{ewt}', \mathsf{lwt})$. Clearly, $\mathcal{A}'$ is a weighted timed automaton over $\mathcal{K}$, Γ and $\mathcal{F}$. Next, we show that $\|\mathcal{A}'\| = \overline{\pi}(\|\mathcal{A}\|)$.

Let $v \in T\Gamma^*$ be of the form $(b_1, t_1)...(b_k, t_k)$ and $\mathcal{R}$ be the set of runs of $\mathcal{A}$ on $w \in T\Sigma^*$ such that $\pi(w) = v$. Let $r, r' \in \mathcal{R}$ be of the form

$$r = (l_0, \nu_0) \xrightarrow{\delta_1} \xrightarrow{e_1} ... \xrightarrow{\delta_{|w|}} \xrightarrow{e_{|w|}} (l_{|w|}, \nu_{|w|})$$

and

$$r' = (l'_0, \nu'_0) \xrightarrow{\delta_1} \xrightarrow{e'_1} ... \xrightarrow{\delta_{|w|}} \xrightarrow{e'_{|w|}} (l'_{|w|}, \nu'_{|w|}).$$

We say that r and r' are equivalent, written $r \equiv r'$, if $l_i = l'_i$ and $\nu_i = \nu'_i$ for $0 \leq i \leq |w|$. Intuitively, $r \equiv r'$ if the runs differ at most in the labels, guards and reset sets of their edges, provided that π maps the labels to the same image and the resulting clock valuations are the same. We use $\mathcal{R}_{/\equiv}$ to denote the set of all equivalence classes induced by $\equiv$. From the fact that $\equiv$ induces a partition of $\mathcal{R}$, we obtain

$$\sum_{\substack{w\in T\Sigma^* \\ \pi(w)=v}} (\|\mathcal{A}\|, w) = \sum_{R\in\mathcal{R}_{/\equiv}} \sum_{r\in R} \mathsf{rwt}(r).$$

Next, let $R \in \mathcal{R}_{/\equiv}$ and $r \in R$ be of the form $(l_0, \nu_0) \xrightarrow{\delta_1} \xrightarrow{e_1} \ldots \xrightarrow{\delta_{|w|}} \xrightarrow{e_{|w|}} (l_{|w|}, \nu_{|w|})$. We define r_R to be the sequence that is obtained from r by replacing $e_i = (l_{i-1}, a_i, \phi_i, \lambda_i, l_i)$ for each $i \in \mathsf{dom}(w)$ by the corresponding edge $e'_i = (l_{i-1}, \pi(a_i), \phi_i, \lambda_i, l_i) \in E'$. We neither change the clock constraints ϕ_i nor the reset sets λ_i, so we have $\nu'_{i-1} \models \phi_i$ and $\nu'_i = (\nu'_{i-1} + \delta_i)[\lambda_i := 0]$ for each $i \in \mathsf{dom}(w)$, and thus, r_R is a run of $\mathcal{A}'$ on v. Moreover, the set of runs of $\mathcal{A}'$ on v is precisely the set of such runs r_R for each $R \in \mathcal{R}_{/\equiv}$, i.e., we have

$$(\|\mathcal{A}'\|, v) = \sum_{R\in\mathcal{R}_{/\equiv}} \mathsf{rwt}(r_R),$$

where r_R is the run of $\mathcal{A}'$ on v obtained from an arbitrary run $r \in \mathcal{R}$ as described above. Next, we show that for every $R \in \mathcal{R}_{/\equiv}$ we have $\mathsf{rwt}(r_R) = \sum_{r\in R} \mathsf{rwt}(r)$, which, with the help of the two equations above, implies the result. Let $R \in \mathcal{R}_{/\equiv}$ and $r \in R$ as above. Then, the following equation holds by distributivity of $\mathcal{K}$:

$$\begin{aligned}
\mathsf{rwt}(r_R) &= \mathsf{in}(l_0) \cdot \left(\prod_{1\le i\le |v|} \mathsf{lwt}(l_{i-1})(\delta_i) \cdot \mathsf{ewt}(e'_i) \right) \cdot \mathsf{out}(l_{|v|}) \\
&= \mathsf{in}(l_0) \cdot \left(\prod_{1\le i\le |v|} \mathsf{lwt}(l_{i-1})(\delta_i) \cdot \sum_{\substack{(l_{i-1},a_i,\phi_i,\lambda_i,l_i)\in E \\ \pi(a_i)=b_i}} \mathsf{ewt}(l_{i-1}, a_i, \phi_i, \lambda_i, l_i) \right) \cdot \mathsf{out}(l_{|v|}) \\
&= \sum_{\substack{(l_{i-1},a_i,\phi_i,\lambda_i,l_i)\in E \\ \pi(a_i)=b_i}} \mathsf{in}(l_0) \cdot \left(\prod_{1\le i\le |v|} \mathsf{lwt}(l_{i-1})(\delta_i) \cdot \mathsf{ewt}(l_{i-1}, a_i, \phi_i, \lambda_i, l_i) \right) \cdot \mathsf{out}(l_{|v|}) \\
&= \sum_{r\in R} \mathsf{rwt}(r).
\end{aligned}$$

Hence, $(\|\mathcal{A}'\|, v) = \sum_{\substack{w\in T\Sigma^* \\ \pi(w)=v}} (\|\mathcal{A}\|, w)$, and thus $\|\mathcal{A}'\| = \bar{\pi}(\|\mathcal{A}\|)$. ■

Lemma 3.8. *The class of $\mathcal{F}$-recognizable timed series is closed under inverse renamings.*

Proof. Let $\pi : \Sigma \to \Gamma$ be a renaming and $\mathcal{T} : T\Gamma^* \to K$ be $\mathcal{F}$-recognizable over Γ. Then there is a weighted timed automaton $\mathcal{A} = (\mathcal{L}, \mathcal{C}, E, \mathsf{in}, \mathsf{out}, \mathsf{ewt}, \mathsf{lwt})$ over $\mathcal{K}$, Γ and $\mathcal{F}$ such that $\|\mathcal{A}\| = \mathcal{T}$. Define $E' = \{(l, a, \phi, \lambda, l') : (l, \pi(a), \phi, \lambda, l') \in \mathcal{L}\}$

and $\mathsf{ewt}'(l, a, \phi, \lambda, l') = \mathsf{ewt}(l, \pi(a), \phi, \lambda, l')$. Then the behaviour of the weighted timed automaton $\mathcal{A}' = (\mathcal{L}, \mathcal{C}, E', \mathsf{in}, \mathsf{out}, \mathsf{ewt}', \mathsf{lwt})$ over $\mathcal{K}$, Σ and $\mathcal{F}$ precisely corresponds to $\bar{\pi}^{-1}(\|\mathcal{A}\|)$. ∎

The next lemma states an important property between TA-recognizable timed languages and $\mathcal{F}$-recognizability of their corresponding characteristic timed series. It will be needed in Sect. 5 and 6.

Lemma 3.9. *Let $\mathbb{1} \in \mathcal{F}$ and $L \subseteq T\Sigma^*$.*

1. *If L is unambiguously TA-recognizable, then 1_L is $\mathcal{F}$-recognizable.*

2. *If $\mathcal{K}$ is idempotent and L is TA-recognizable, then 1_L is $\mathcal{F}$-recognizable over $\mathcal{K}$.*

PROOF. 1.: Let $L \subseteq T\Sigma^*$ be unambiguously TA-recognizable. Then there is an unambiguous timed automaton $\mathcal{A} = (\mathcal{L}, \mathcal{L}_0, \mathcal{L}_f, \mathcal{C}, E)$ over Σ such that $L(\mathcal{A}) = L$. Let $\mathcal{A}' = (\mathcal{L}, \mathcal{C}, E, \mathsf{in}, \mathsf{out}, \mathsf{ewt}, \mathsf{lwt})$ be the weighted timed automaton over $\mathcal{K}$, Σ and $\mathcal{F}$ obtained from $\mathcal{A}$ as follows: define $\mathsf{in}(l) = 1$ if $l \in \mathcal{L}_0$ and $\mathsf{in}(l) = 0$ otherwise; and $\mathsf{out}(l) = 1$ if $l \in \mathcal{L}_f$ and $\mathsf{out}(l) = 0$ otherwise for each $l \in \mathcal{L}$, $\mathsf{ewt}(e) = 1$ for each $e \in E$ and $\mathsf{lwt}(l) = \mathbb{1}$ for every $l \in \mathcal{L}$ in $\mathcal{A}$. One can easily see that $\|\mathcal{A}'\| = 1_L$.

2.: The proof can be done analogously to 1. Using the fact that $\mathcal{K}$ is idempotent and hence $1 + 1 = 1$, we do not need unambiguity of $\mathcal{A}$. ∎

4 A Kleene-Schützenberger Theorem for Weighted Timed Automata

4.1 Introduction

The goal of this section is to provide a characterization of the behaviour of weighted timed automata in terms of *rational* timed series, i.e., timed series constructed by the standard rational operations sum, Cauchy product and Kleene star iteration. In the theory of formal languages, rational expressions are an important formalism to specify the behaviour of finite systems. Here, we lift this formalism to the weighted timed setting. This provides a new tool for the specification of the behaviour of real-time systems. The main result is a timed analogue of the Schützenberger theorem [101], i.e., we show that the formalism of rational timed series is equivalent to the expressive power of weighted timed automata. The theorem we present here is close to the corresponding theorems in the frameworks of formal language theory and the theory of WFA-recognizable series. The translations from weighted timed automata to rational timed series and vice versa are done similarly to the classical case, but differ in some important details.

For the results presented in this section, we extend the work on a Kleene theorem for timed automata. There have been several approaches to give a Kleene theorem for TA-recognizable timed languages [12, 14, 13, 28, 59, 60]. However, we choose the latest approach of Bouyer and Petit [29] because of its simplicity and elegance. According to their result, the class of TA-recognizable timed languages coincides with the class of rational timed languages, defined using the operations sum, concatenation, Kleene star iteration and an additional projection operation. For the proof of our main result, we combine the methods of Bouyer and Petit, Schützenberger and new techniques, summarized in the following.

We follow the approach of Bouyer and Petit [29] and define the semantics of weighted timed automata in a slightly different manner as introduced in Sect. 2.1, based on the notion of *clock words*. Clock words, as opposed to timed words [6], bear information concerning the actual values of clock variables. In this way, they enable us to define a concatenation operation (and the induced iteration operation) in a natural way and close to the concatenation operation known from the classical theory of formal languages. Consequently, all the definitions and constructions for our Kleene-Schützenberger theorem are given with respect to clock words. Then, since clock words can easily be mapped to timed words using a projection, the theorem for clock words can be extended to timed words. To bring weights into play, we introduce the notion of *clock series*, a special kind of series mapping clock words to elements in the semiring. We define sum, Cauchy product and (finite) Kleene star iteration on the class of clock series and present a formal definition for the class of rational clock series. The main objective of this section is to show that this class is equal to the class of recognizable clock series. We establish this in two steps. First, we prove that the class of recognizable clock series is closed under the three operations mentioned above. In our proof, for dealing with the weight functions assigned to locations, we need to give new methods for normalizing weighted timed automata. It follows that any rational clock series is recognizable. The proof for the other direction, i.e., that any recognizable clock series is rational, is based on the solution of equations [29, 19], i.e., we present a translation procedure from weighted timed automata to a system of equations, the (unique) solution of which corresponds to a rational clock series. Finally, we show how we can extend the Kleene-Schützenberger theorem in such a way that it can as well be applied to the timed semantics presented in Sect. 2.1.

4.2 Clock Series

In the following, we fix a set $\mathcal{C} = \{x_1, ..., x_n\}$ of clock variables, a semiring $\mathcal{K}$, a finite alphabet Σ and a family $\mathcal{F}$ of weight functions from $\mathbb{R}_{\geq 0}$ to K.

An *n-clock word* is a finite sequence $w = (t_0, \nu_0)(a_1, t_1, \nu_1)...(a_k, t_k, \nu_k)$ in $(\mathbb{R}_{\geq 0} \times \mathbb{R}^n_{\geq 0})(\Sigma \times \mathbb{R}_{\geq 0} \times \mathbb{R}^n_{\geq 0})^*$, where $(a_1, t_1)...(a_k, t_k)$ is a timed word, and ν_i gives the values of the clock variables just after the computation of a_i. The pair (t_0, ν_0) corresponds to the starting condition and is considered to be an *empty n-clock word.* We write $C_n\Sigma^*$ for the set of n-clock words over Σ. The set $\mathbb{R}_{\geq 0} \times \mathbb{R}^n_{\geq 0}$ of empty n-clock words is denoted by $\mathcal{E}_n$. We further define $C_n\Sigma^+$ by $C_n\Sigma^* \backslash \mathcal{E}_n$. Let $w = (t_0, \nu_0)(a_1, t_1, \nu_1)...(a_k, t_k, \nu_k)$ and $w' = (t'_0, \nu'_0)(a'_1, t'_1, \nu'_1)...(a'_l, t'_l, \nu'_l)$ be two n-clock words for $k, l \in \mathbb{N}$. We say that w is *compatible* with w' if $(t_k, \nu_k) = (t'_0, \nu'_0)$. In this case, we define the *concatenation* $w; w'$ of w and w' to be the n-clock word $(t_0, \nu_0)(a_1, t_1, \nu_1)...(a_k, t_k, \nu_k)(a'_1, t'_1, \nu'_1)...(a'_l, t'_l, \nu'_l)$. A function $S : C_n\Sigma^* \to K$ is called an n-clock series. We use $\mathcal{K}\langle\langle C_n\Sigma^* \rangle\rangle$ to denote the class of all n-clock series over $\mathcal{K}$ and Σ. Let $\mathcal{A} = (\mathcal{L}, \mathcal{C}, E, \mathsf{in}, \mathsf{out}, \mathsf{ewt}, \mathsf{lwt})$ be a weighted timed automaton over $\mathcal{K}$, Σ and $\mathcal{F}$. The semantics of $\mathcal{A}$ can be given in terms of n-clock words. This clock semantics is very similar to the timed semantics presented in Sect. 2.1. The set of clock states consists of triples $(l, t, \nu) \in \mathcal{L} \times \mathbb{R}_{\geq 0} \times (\mathbb{R}_{\geq 0})^n$. Between these states, we distinguish between two kinds of transitions, namely *timed clock transition* of the form $(l, t, \nu) \xrightarrow{\delta} (l, t + \delta, \nu + \delta)$ for some $\delta \in \mathbb{R}_{\geq 0}$, and *discrete clock transition* of the form $(l, t, \nu) \xrightarrow{e} (l', t, \nu')$ for some $e = (l, a, \phi, \lambda, l') \in E$ such that $\nu \models \phi$ and $\nu' = (\nu + \delta)[\lambda := 0]$. Again, we write $(l, t, \nu) \xrightarrow{\delta}\xrightarrow{e} (l', t', \nu')$ for a timed clock transition followed by a discrete clock transition and call this a *clock transition.* Let $w = (t_0, \nu_0)(a_1, t_1, \nu_1)...(a_k, t_k, \nu_k)$ be an n-clock word. A *clock run* of $\mathcal{A}$ on w is of the form $(l_0, t_0, \nu_0) \xrightarrow{\delta_1}\xrightarrow{e_1} ... \xrightarrow{\delta_k}\xrightarrow{e_k} (l_k, t_k, \nu_k)$, where $(l_{i-1}, t_{i-1}, \nu_{i-1}) \xrightarrow{\delta_i}\xrightarrow{e_i} (l_i, t_i, \nu_i)$ is a clock transition for each $i \in \{1, ..., k\}$. The running weight $\mathsf{rwt}(r)$ of a clock run r is defined in the same manner as for ordinary runs, i.e., by $\mathsf{rwt}(r) = \mathsf{in}(l_0) \cdot \left(\prod_{1 \leq i \leq k} \mathsf{lwt}(l_{i-1})(\delta_i) \cdot \mathsf{ewt}(e_i)\right) \cdot \mathsf{out}(l_k)$.

The *clock behaviour* $\|\mathcal{A}\|_n$ of $\mathcal{A}$ is the n-clock series $\|\mathcal{A}\|_n : C_n\Sigma^* \to K$ defined by $(\|\mathcal{A}\|_n, w) = \sum\{\mathsf{rwt}(r) : r \text{ is a clock run of } \mathcal{A} \text{ on } w\}$ for each $w \in C_n\Sigma^*$. An n-clock series S is said to be $\mathcal{F}$-recognizable over $\mathcal{K}$ and Σ if there is a weighted timed automaton $\mathcal{A}$ over $\mathcal{K}$, Σ and $\mathcal{F}$ with n clock variables such that $\|\mathcal{A}\|_n = S$. We may omit $\mathcal{K}$ or Σ if they are clear from the context. We use $\mathcal{K}^{\mathcal{F}-rec}\langle\langle C_n\Sigma^*\rangle\rangle$ to denote the class of all $\mathcal{F}$-recognizable n-clock series.

Let $S_1, S_2 \in \mathcal{K}\langle\langle C_n\Sigma^*\rangle\rangle$ and $w \in C_n\Sigma^*$. Similarly to the untimed theory of series, we define the *sum* $S_1 + S_2$ pointwise, i.e., we let $(S_1 + S_2, w) = (S_1, w) + (S_2, w)$. The *Cauchy product* $S_1; S_2$ is defined by $(S_1; S_2, w) = \sum_{u;v=w} (S_1, u) \cdot (S_2, v)$. Furthermore, we define for every $k \in K$ the n-clock series $k\varepsilon : C_n\Sigma^* \to K$ by $(k\varepsilon, w) = k$ if $w \in \mathcal{E}_n$, $(k\varepsilon, w) = 0$ otherwise. We call n-clock series of this form *monomials*. The following lemma is the n-clock series version of the well known fact that the set of series over the free monoid together with sum and Cauchy product is a semiring. The proof proceeds by elementary calculations.

Lemma 4.1. *The structure* $(\mathcal{K}\langle\langle C_n\Sigma^*\rangle\rangle, +, ;, 0\varepsilon, 1\varepsilon)$ *is a semiring.*

For an n-clock series $S \in \mathcal{K}\langle\langle C_n\Sigma^*\rangle\rangle$, let $S^0 = 1\varepsilon$ and, inductively, $S^k = S; S^{k-1}$ be the *k-th power* of S for $k \geq 1$. The n-clock series S is called *proper* if $(S, w) = 0$ for each $w \in \mathcal{E}_n$. For proper n-clock series S, we define the *Kleene star iteration* S^* by

$$(S^*, w) = \sum_{k \geq 0} (S^k, w).$$

Notice that from $(S, w) = 0$ for $w \in \mathcal{E}_n$, it follows that $(S^k, w) = 0$ for any $k > |w|$. This implies that the sum given above is finite and hence exists in $\mathcal{K}$. Straightforward calculations prove the following Lemma.

Lemma 4.2. *Let $S : C_n\Sigma^* \to K$ be a proper n-clock series. Then $S; S^* + 1\varepsilon = S^*$.*

The next lemma is crucial in Sect. 4.4.

Lemma 4.3. *Let $S_1, S_2 : C_n\Sigma^* \to K$ be n-clock series and let S_1 be proper. Then the equation $S = S_1; S + S_2$ for some n-clock series $S : C_n\Sigma^* \to K$ has the unique solution $S_1^*; S_2$.*

Proof. First, we show that $S_1^*; S_2$ is a solution. By Lemmas 4.1 and 4.2 we have

$$S_1; (S_1^*; S_2) + S_2 = ((S_1; S_1^*); S_2) + S_2 = (S_1; S_1^* + 1\varepsilon); S_2 = S_1^*; S_2.$$

For uniqueness, let S_{sol} be an arbitrary solution of the equation. By substitution and Lemma 4.1, we can show that $S_{\text{sol}} = S_1^k; S_{\text{sol}} + (\sum_{0 \leq i < k} S_1^i); S_2$ for each $k \in \mathbb{N}$. Thus, for any $w \in C_n\Sigma^*$ we obtain

$$\begin{aligned}(S_{\text{sol}}, w) &= (S_1^{|w|+1}; S_{\text{sol}}, w) + ((\sum_{0 \leq i < |w|+1} S_1^i); S_2, w) \\ &\overset{\star}{=} 0 + (S_1^*; S_2, w) \\ &= (S_1^*; S_2, w).\end{aligned}$$

At * we use that S_1 is proper and hence $(S_1^i, w) = 0$ for each $i > |w|$. Hence, $S_1^*; S_2$ is the unique solution. ■

Next, we give an explicit formula for the calculation of S^k. It can be proved by induction on k.

Lemma 4.4. *If $S : C_n\Sigma^* \to K$ is a proper n-clock series, $k \in \mathbb{N}$ and $w \in C_n\Sigma^*$, then (S^k, w) has the explicit representation*

$$(S^k, w) = \sum_{w = w_1; \ldots; w_k} \prod_{i=1}^{k} (S, w_i).$$

For $f \in \mathcal{F}$, $k \in K$, $a \in \Sigma$, $\phi \in \Phi(\mathcal{C})$, and $\lambda \subseteq \mathcal{C}$, we define the $\mathcal{F}$-monomial $\langle f, k, a, \phi, \lambda \rangle : C_n\Sigma^* \to K$ as follows:

$$(\langle f, k, a, \phi, \lambda \rangle, w) = \begin{cases} f(\delta) \cdot k & \text{if } w = (t, \nu)(a, t+\delta, \nu') \in C_n\Sigma^* \text{ for some } \delta, t \in \mathbb{R}_{\geq 0}, \\ & \nu \in \mathbb{R}^n_{\geq 0} \text{ such that } \nu + \delta \models \phi \text{ and } \nu' = (\nu + \delta)[\lambda := 0], \\ 0 & \text{otherwise.} \end{cases}$$

Now we give the important notion of rationality for n-clock series. An n-clock series S is *$\mathcal{F}$-rational over $\mathcal{K}$ and Σ* if it can be defined starting from finitely many monomials and $\mathcal{F}$-monomials, by means of a finite number of applications of $+$, ; and *, where the latter may only be applied to proper n-clock series.

Example 4.5. Consider the following specification of a real-time system with a single resource, where $\Gamma = \{a, b, c, d\}$ is a set of actions:

> The system must execute a and b, and b must be executed exactly 3 time units after a. Between a and b, action c (costs €3) and action d (costs €2) may be executed consecutively for an arbitrary number of times, but d is restricted to happen strictly between 1 and 2 time units after c. Being in the state after action a or d has been executed, costs €5 per time unit, whereas being in the state after c has been executed, costs €1 per time unit.

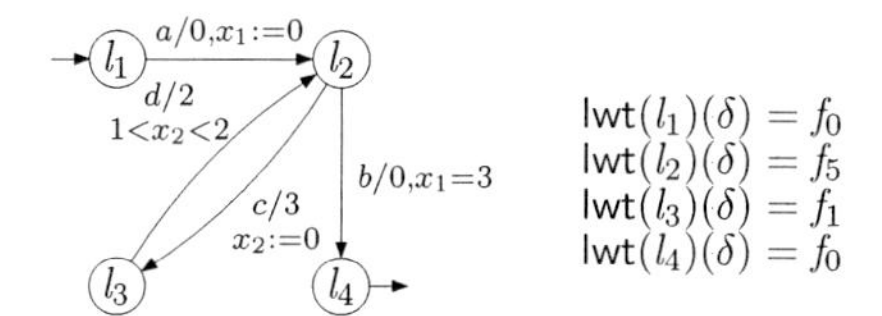

Figure 4.1: The weighted timed automaton for Example 4.5

The specification can be represented by the following $\mathcal{F}$-rational clock series over the min-plus-semiring and Γ, where $\mathcal{F}$ is the family of linear functions $f_i(\delta)$ defined by $f_i(\delta) = i \cdot \delta$ for each $i, \delta \in \mathbb{R}_{\geq 0}$:

$$\langle f_0, 0, a, \top, \{x_1\}\rangle \big(\langle f_5, 3, c, \top, \{x_2\}\rangle\langle f_1, 2, d, 1 < x_2 < 2, \emptyset\rangle\big)^* \langle f_5, 0, b, x_1 = 3, \emptyset\rangle$$

where $\top$ means *true*. In Fig. 4.2, we give the corresponding weighted timed automaton.

Observe that, similarly to the case of weighted timed automata, by specifying $\mathcal{K}$ and $\mathcal{F}$, we obtain $\mathcal{F}$-rational expressions for several other automata classes. In particular, if $\mathcal{K}$ is the Boolean semiring and $\mathcal{F}$ is the family of constant functions $\mathbb{1}$, then $\mathcal{F}$-rational n-clock series correspond to rational n-clock expressions as defined by Bouyer and Petit [29].

4.3 From Rationality to Recognizability

In this section, we prove one inclusion of our Kleene-Schützenberger theorem for the class of weighted timed automata, namely that each $\mathcal{F}$-rational n-clock series is $\mathcal{F}$-recognizable. To this end, we define weighted timed automata recognizing the basic n-clock series, namely monomials and $\mathcal{F}$-monomials, respectively. Then, we present new normalization techniques that will be used to finally show that the class of $\mathcal{F}$-recognizable n-clock series is closed under sum, the Cauchy product and Kleene star iteration.

Lemma 4.6. *Let $k \in K$. Then the monomial $k\varepsilon$ is $\mathcal{F}$-recognizable.*

PROOF. Let $k \in K$. The clock behaviour of the weighted timed automaton $\mathcal{A} = (\mathcal{L}, \mathcal{C}, E, \mathsf{in}, \mathsf{out}, \mathsf{ewt}, \mathsf{lwt})$, where $\mathcal{L} = \{l\}$, $E = \emptyset$, $\mathsf{in}(l) = k$, $\mathsf{out}(l) = 1$ and $\mathsf{lwt}(l) \in \mathcal{F}$, corresponds to $k\varepsilon$. ■

Lemma 4.7. *Let $f \in \mathcal{F}$, $k \in K$, $a \in \Sigma$, $\phi \in \Phi(\mathcal{C})$ and $\lambda \subseteq \mathcal{C}$. Then the $\mathcal{F}$-monomial $\langle f, k, a, \phi, \lambda\rangle$ is $\mathcal{F}$-recognizable.*

PROOF. Let $f \in \mathcal{F}$, $k \in K$, $a \in \Sigma$, $\phi \in \Phi(\mathcal{C})$ and $\lambda \subseteq \mathcal{C}$. We define the weighted timed automaton $\mathcal{A}_{\langle f,k,a,\phi,\lambda\rangle} = (\mathcal{L}, \mathcal{C}, E, \mathsf{in}, \mathsf{out}, \mathsf{ewt}, \mathsf{lwt})$, where

- $\mathcal{L} = \{l_1, l_2\}$,
- $E = \{(l_1, a, \phi, \lambda, l_2)\}$,
- $\mathsf{in}(l_1) = 1$, $\mathsf{in}(l_2) = 0$,
- $\mathsf{out}(l_1) = 0$, $\mathsf{out}(l_2) = 1$,
- $\mathsf{ewt}(l_1, a, \phi, \lambda, l_2) = k$,
- $\mathsf{lwt}(l_1) = f$, $\mathsf{lwt}(l_2) \in \mathcal{F}$.

Clearly, $\|\mathcal{A}_{\langle f,k,a,\phi,\lambda\rangle}\|_n = \langle f, k, a, \phi, \lambda\rangle$. ■

For the proof of closure of the class of recognizable n-clock series under sum, we refer to the corresponding proof for recognizable timed series in Sect. 3. It can be easily adapted to the case of n-clock series and we obtain the following lemma.

Lemma 4.8. *The class of $\mathcal{F}$-recognizable n-clock series is closed under sum.*

In the following, we give normalization techniques for weighted timed automata which will be essential for subsequent lemmas. For showing closure of the class of $\mathcal{F}$-recognizable n-clock series under the Cauchy product, we need the weighted timed automaton to be final-location-normalized. We say that a weighted timed automaton $\mathcal{A}$ is *final-location-normalized* if there is a location l_f such that $\mathsf{out}(l) = 1$ if $l = l_f$ and $\mathsf{out}(l) = 0$ otherwise, $\mathsf{in}(l_f) = 0$, and l_f has no outgoing edges. We call l_f the *sink location* of $\mathcal{A}$.

Lemma 4.9. *For each weighted timed automaton $\mathcal{A}$, there is a final-location-normalized weighted timed automaton $\mathcal{A}'$ such that $(\|\mathcal{A}'\|_n, w) = (\|\mathcal{A}\|_n, w)$ for each $w \in C_n\Sigma^+$ and $(\|\mathcal{A}'\|_n, w) = 0$ for each $w \in \mathcal{E}_n$.*

Proof. Let $\mathcal{A} = (\mathcal{L}, \mathcal{C}, E, \mathsf{in}, \mathsf{out}, \mathsf{ewt}, \mathsf{lwt})$ be a weighted timed automaton over $\mathcal{K}$, Σ and $\mathcal{F}$. Define $\mathcal{A}' = (\mathcal{L}', \mathcal{C}, E', \mathsf{in}', \mathsf{out}', \mathsf{ewt}', \mathsf{lwt}')$, where

- $\mathcal{L}' = \mathcal{L} \cup \{l_f\}$,
- $\mathsf{in}'(l) = \begin{cases} \mathsf{in}(l) & \text{if } l \in \mathcal{L} \\ 0 & \text{if } l = l_f, \end{cases}$
- $\mathsf{out}'(l) = \begin{cases} 1 & \text{if } l = l_f \\ 0 & \text{otherwise}, \end{cases}$
- $E' = E \cup \{(l, a, \phi, \lambda, l_f) : (l, a, \phi, \lambda, l') \in E\}$,

- $\mathsf{ewt}'(e) = \mathsf{ewt}(e)$ if $e \in E$, $\mathsf{ewt}'(l, a, \phi, \lambda, l_f) = \sum_{(l,a,\phi,\lambda,l') \in E} \mathsf{ewt}(l, a, \phi, \lambda, l') \cdot \mathsf{out}(l')$,
- $\mathsf{lwt}'(l) = \mathsf{lwt}(l)$ if $l \in \mathcal{L}$ and $\mathsf{lwt}'(l_f) \in \mathcal{F}$.

Intuitively, we redirect all edges to the new final location. The weight of each of these new edges must be the sum of the weights of all "equivalent" edges, i.e., edges with the same label, clock constraint and reset set, multiplied with the weight for leaving the destination location of these edges. Using this notion of equivalence, we show that $(\|\mathcal{A}'\|_n, w) = (\|\mathcal{A}\|_n, w)$ for any $w \in C_n\Sigma^+$.

Let $w = (t_0, \nu_0)(a_1, t_1, \nu_1)...(a_k, t_k, \nu_k) \in C_n\Sigma^+$. We let $\mathcal{R}$ be the set of clock runs of $\mathcal{A}$ on w. Consider two clock clock runs $r, r' \in \mathcal{R}$

$$r = (l_0, t_0, \nu_0) \xrightarrow{\delta_1} \xrightarrow{e_1} (l_1, t_1, \nu_1) \xrightarrow{\delta_2} \xrightarrow{e_2} ... \xrightarrow{\delta_k} \xrightarrow{e_k} (l_k, t_k, \nu_k)$$

and

$$r' = (l_0', t_0, \nu_0) \xrightarrow{\delta_1} \xrightarrow{e_1'} (l_1', t_1, \nu_1) \xrightarrow{\delta_2} \xrightarrow{e_2'} ... \xrightarrow{\delta_k} \xrightarrow{e_k'} (l_k', t_k, \nu_k).$$

We define an equivalence relation $\equiv$ on $\mathcal{R}$ as follows: $r \equiv r'$ if and only if r and r' only differ in the last edge, i.e., we have $e_i = e_i'$ for each $i \in \{1, ..., k-1\}$. We denote the set of induced equivalence classes by $\mathcal{R}_{/\equiv}$. The equivalence relation $\equiv$ induces a partition on $\mathcal{R}$. This implies

$$(\|\mathcal{A}\|_n, w) = \sum_{R \in \mathcal{R}_{/\equiv}} \sum_{r \in R} \mathsf{rwt}(r).$$

Now, let $R \in \mathcal{R}_{/\equiv}$. We denote by r^R the clock run of $\mathcal{A}'$ that results from an arbitrary clock run $r \in R$ of the form

$$r = (l_0, t_0, \nu_0) \xrightarrow{\delta_1} \xrightarrow{e_1} (l_1, t_1, \nu_1) \xrightarrow{\delta_2} \xrightarrow{e_2} ... \xrightarrow{\delta_k} \xrightarrow{e_k} (l_k, t_k, \nu_k)$$

by replacing the edge e_k by the corresponding edge in E', i.e., if $e_k = (l_{k-1}, a_k, \phi_k, \lambda_k, l_k)$, then we replace it by the edge $(l_{k-1}, a_k, \phi_k, \lambda_k, l_f)$. One can easily see that the set of clock runs of $\mathcal{A}'$ on w with $l_{|w|}$ satisfying $\mathsf{out}(l_{|w|}) \neq 0$ is exactly the set of such clock runs r^R for each $R \in \mathcal{R}_{/\equiv}$. We obtain $(\|\mathcal{A}'\|_n, w) = \sum_{R \in \mathcal{R}_{/\equiv}} \mathsf{rwt}(r^R)$. By the two equalities above, in order to show $(\|\mathcal{A}'\|_n, w) = (\|\mathcal{A}\|_n, w)$, it suffices to show

$$\mathsf{rwt}(r^R) = \sum_{r \in R} \mathsf{rwt}(r)$$

for each $R \in \mathcal{R}_{/\equiv}$. So let $R \in \mathcal{R}_{/\equiv}$. For every $r \in R$, we can write the running weight of r as follows:

$$\mathsf{rwt}(r) \quad = \quad \mathsf{in}(l_0) \cdot \left(\prod_{1 \leq i \leq k} \mathsf{lwt}(l_{i-1})(\delta_i) \cdot \mathsf{ewt}(e_i) \right) \cdot \mathsf{out}(l_k)$$

$$= \mathsf{in}(l_0) \cdot \left(\prod_{1 \le i < k} \mathsf{lwt}(l_{i-1})(\delta_i) \cdot \mathsf{ewt}(e_i) \right) \cdot \mathsf{lwt}(l_{k-1})(\delta_k) \cdot \mathsf{ewt}(e_k) \cdot \mathsf{out}(l_k).$$

Thus, we can write

$$\sum_{r \in R} \mathsf{rwt}(r) = \sum_{r \in R} \mathsf{in}(l_0) \cdot \left(\prod_{1 \le i < k} \mathsf{lwt}(l_{i-1})(\delta_i) \cdot \mathsf{ewt}(e_i) \right) \cdot \mathsf{lwt}(l_{k-1})(\delta_k) \cdot \mathsf{ewt}(e_k) \cdot \mathsf{out}(l_k).$$

The term $\mathsf{in}(l_0) \cdot \left(\prod_{1 \le i < k} \mathsf{lwt}(l_{i-1})(\delta_i) \cdot \mathsf{ewt}(e_i) \right) \cdot \mathsf{lwt}(l_{k-1})(\delta_k)$ is the same for each $r \in R$ due to the definition of $\equiv$. Using distributivity, we obtain

$$\begin{aligned} &\sum_{r \in R} \mathsf{rwt}(r) \\ = \;& \mathsf{in}(l_0) \cdot \left(\prod_{1 \le i < k} \mathsf{lwt}(l_{i-1})(\delta_i) \cdot \mathsf{ewt}(e_i) \right) \cdot \mathsf{lwt}(l_{k-1})(\delta_k) \cdot \sum_{r \in R} \mathsf{ewt}(e_k) \cdot \mathsf{out}(l_k) \\ = \;& \mathsf{in}(l_0) \cdot \left(\prod_{1 \le i < k} \mathsf{lwt}(l_{i-1})(\delta_i) \cdot \mathsf{ewt}(e_i) \right) \cdot \mathsf{lwt}(l_{k-1})(\delta_k) \\ & \cdot \sum_{(l_{k-1}, a_k, \phi_k, \lambda_k, l_k) \in E} \mathsf{ewt}(l_{k-1}, a_k, \phi_k, \lambda_k, l_k) \cdot \mathsf{out}(l_k). \end{aligned}$$

Now, consider the clock run r^R. Using the definition of clock runs of this form from above, as well as the definition of the weight functions in $\mathcal{A}'$, we get

$$\begin{aligned} &\mathsf{rwt}(r^R) \\ = \;& \mathsf{in}(l_0) \cdot \left(\prod_{1 \le i < k} \mathsf{lwt}(l_{i-1})(\delta_i) \cdot \mathsf{ewt}(e_i) \right) \cdot \mathsf{lwt}(l_{k-1})(\delta_k) \cdot \\ & \cdot \sum_{(l_{k-1}, a_k, \phi_k, \lambda_k, l_k) \in E} \mathsf{ewt}(l_{k-1}, a_k, \phi_k, \lambda_k, l_k) \cdot \mathsf{out}(l_k). \end{aligned}$$

Hence, we have shown that $(\|\mathcal{A}'\|_n, w) = (\|\mathcal{A}\|_n, w)$ for each $w \in C_n\Sigma^+$.

Finally, let $w \in \mathcal{E}_n$. Then we have $(\|\mathcal{A}'\|_n, w) = \sum_{l \in \mathcal{L}} \mathsf{in}(l) \cdot \mathsf{out}(l) = 0$, since for each $l \in \mathcal{L}'$ we have $\mathsf{in}(l) \neq 0$ if and only if $\mathsf{out}(l) = 0$. This finishes the proof. ■

Lemma 4.10. *The class of $\mathcal{F}$-recognizable n-clock series is closed under Cauchy products.*

PROOF. Let $S_1, S_2 : C_n\Sigma^* \to K$ be two $\mathcal{F}$-recognizable n-clock series. Let $w_0 \in \mathcal{E}_n$. Further let $s = (S_1, w_0)$ and write $S_1 = S_1' + s\varepsilon$, where S_1' is proper. Similarly, let $s' = (S_2, w_0)$ and write $S_2 = S_2' + s'\varepsilon$, where S_2' is proper. By Lemma 4.1, we have $S_1; S_2 = (S_1'; S_2') + (s\varepsilon; S_2') + (S_1'; s'\varepsilon) + (s\varepsilon; s'\varepsilon)$. In the following, we show that all summands are $\mathcal{F}$-recognizable. Then, by Lemma 4.8, $S_1; S_2$ is $\mathcal{F}$-recognizable, too.

By Lemma 4.9, there is a final-location-normalized weighted timed automaton $\mathcal{A}_1 = (\mathcal{L}_1, \mathcal{C}, E_1, \mathsf{in}_1, \mathsf{out}_1, \mathsf{ewt}_1, \mathsf{lwt}_1)$ such that $\|\mathcal{A}_1\|_n = S_1'$. We use l_f to denote the sink location of $\mathcal{A}_1$. Further, there is a final-location-normalized weighted timed automaton $\mathcal{A}_2 = (\mathcal{L}_2, \mathcal{C}, E_2, \mathsf{in}_2, \mathsf{out}_2, \mathsf{ewt}_2, \mathsf{lwt}_2)$ over $\mathcal{K}$, Σ and $\mathcal{F}$ such that $\|\mathcal{A}_2\|_n = S_2'$. We may assume $\mathcal{L}_1 \cap \mathcal{L}_2 = \emptyset$.

First, we show that $S_1'; S_2'$ is $\mathcal{F}$-recognizable. We define $\mathcal{A} = (\mathcal{L}, \mathcal{C}, E, \mathsf{in}, \mathsf{out}, \mathsf{ewt}, \mathsf{lwt})$, where

- $\mathcal{L} = \mathcal{L}_1 \cup \mathcal{L}_2$,
- $E = E_1 \cup E_2 \cup \{(l, a, \phi, \lambda, l') : l' \in \mathcal{L}_2 \text{ such that } \mathsf{in}_2(l') \neq 0, (l, a, \phi, \lambda, l_f) \in E_1\}$,
- $\mathsf{in}(l) = \begin{cases} \mathsf{in}_1(l) & \text{if } l \in \mathcal{L}_1 \\ 0 & \text{otherwise,} \end{cases}$
- $\mathsf{out}(l) = \begin{cases} 0 & \text{if } l \in \mathcal{L}_1 \\ \mathsf{out}_2(l) & \text{otherwise,} \end{cases}$
- $\mathsf{ewt}(l, a, \phi, \lambda, l') = \begin{cases} \mathsf{ewt}_1(l, a, \phi, \lambda, l') & \text{if } (l, a, \phi, \lambda, l') \in E_1 \\ \mathsf{ewt}_2(l, a, \phi, \lambda, l') & \text{if } (l, a, \phi, \lambda, l') \in E_2 \\ \mathsf{ewt}_1(l, a, \phi, \lambda, l_f) \cdot \mathsf{in}_2(l') & \text{otherwise,} \end{cases}$
- $\mathsf{lwt}(l) = \begin{cases} \mathsf{lwt}_1(l) & \text{if } l \in \mathcal{L}_1 \\ \mathsf{lwt}_2(l) & \text{otherwise.} \end{cases}$

We show that $\|\mathcal{A}\|_n = \|\mathcal{A}_1\|_n; \|\mathcal{A}_2\|_n$. First note that $\|\mathcal{A}\|_n$ is proper and thus $(\|\mathcal{A}\|_n, w) = (\|\mathcal{A}_1\|_n; \|\mathcal{A}_2\|_n, w) = 0$ for each $w \in \mathcal{E}_n$. So let $w \in C_n\Sigma^+$. We further observe that each clock run r of $\mathcal{A}$ on w with $\mathsf{rwt}(r) \neq 0$ must contain exactly one edge $e \in E \backslash (E_1 \cup E_2)$. Each such clock run can be uniquely decomposed into clock runs r_1 of $\mathcal{A}_1$ on $w_1 \in C_n\Sigma^+$ and r_2 of $\mathcal{A}_2$ on $w_2 \in C_n\Sigma^+$ such that $w_1; w_2 = w$ and $\mathsf{rwt}(r) = \mathsf{rwt}(r_1) \cdot \mathsf{rwt}(r_2)$, as is shown in the following.

Let $w = (t_0, \nu_0)(a_1, t_1, \nu_1)...(a_m, t_m, \nu_m)$ and $r = (l_0, t_0, \nu_0) \xrightarrow{\delta_1} \xrightarrow{e_1} ... \xrightarrow{\delta_m} \xrightarrow{e_m} (l_m, t_m, \nu_m)$ be a clock run of $\mathcal{A}$ on w and assume there is some $k \in \{1, ..., m\}$ such that $e_k \in E \backslash (E_1 \cup E_2)$. Hence there is some edge $e = (l_{k-1}, a_k, \phi_k, \lambda_k, l_f) \in E_1$. We split r after the k-th edge and use e instead of e_k in the first part of r, obtaining the clock run $r_1 = (l_0, t_0, \nu_0) \xrightarrow{\delta_1} \xrightarrow{e_1} ... \xrightarrow{\delta_k} \xrightarrow{e} (l_f, t_k, \nu_k)$ of $\mathcal{A}_1$ on $w_1 = (t_0, \nu_0)(a_1, t_1, \nu_1)...(a_k, t_k, \nu_k) \in$

$C_n\Sigma^+$ and the clock run $r_2 = (l_k, t_k, \nu_k) \xrightarrow{\delta_{k+1}}\xrightarrow{e_{k+1}} \ldots \xrightarrow{\delta_m}\xrightarrow{e_m} (l_m, t_m, \nu_m)$ of $\mathcal{A}_2$ on $w_2 = (t_k, \nu_k)(a_{k+1}, t_{k+1}, \nu_{k+1})\ldots(a_m, t_m, \nu_m) \in C_n\Sigma^+$. Finally, we have

$$\begin{aligned}
& \mathsf{rwt}(r) \\
= \; & \mathsf{in}(l_0) \cdot \left(\prod_{1 \le i \le m} \mathsf{lwt}(l_{i-1})(\delta_i) \cdot \mathsf{ewt}(e_i) \right) \cdot \mathsf{out}(l_m) \\
= \; & \mathsf{in}(l_0) \cdot \left(\prod_{1 \le i < k} \mathsf{lwt}(l_{i-1})(\delta_i) \cdot \mathsf{ewt}(e_i) \right) \cdot \mathsf{lwt}(l_{k-1})(\delta_k) \cdot \mathsf{ewt}(e_k) \\
& \cdot \left(\prod_{k < i \le m} \mathsf{lwt}(l_{i-1})(\delta_i) \cdot \mathsf{ewt}(e_i) \right) \cdot \mathsf{out}(l_m) \\
\overset{\star}{=} \; & \mathsf{in}_1(l_0) \cdot \left(\prod_{1 \le i < k} \mathsf{lwt}_1(l_{i-1})(\delta_i) \cdot \mathsf{ewt}_1(e_i) \right) \cdot \mathsf{lwt}_{k-1}(\delta_k) \cdot \mathsf{ewt}_1(e) \\
& \cdot \mathsf{in}_2(l_k) \cdot \left(\prod_{k < i \le m} \mathsf{lwt}_2(l_{i-1})(\delta_i) \cdot \mathsf{ewt}_2(e_i) \right) \cdot \mathsf{out}_2(l_m) \\
\overset{\star\star}{=} \; & \mathsf{in}_1(l_0) \cdot \left(\prod_{1 \le i < k} \mathsf{lwt}_1(l_{i-1})(\delta_i) \cdot \mathsf{ewt}_1(e_i) \right) \cdot \mathsf{lwt}_1(l_{k-1})(\delta_k) \cdot \mathsf{ewt}_1(e) \cdot \mathsf{out}_1(l_f) \\
& \cdot \mathsf{in}_2(l_k) \cdot \left(\prod_{k < i \le m} \mathsf{lwt}_2(l_{i-1})(\delta_i) \cdot \mathsf{ewt}_2(e_i) \right) \cdot \mathsf{out}_2(l_m) \\
= \; & \mathsf{rwt}(r_1) \cdot \mathsf{rwt}(r_2)
\end{aligned}$$

where at * we use $\mathsf{ewt}(e_k) = \mathsf{ewt}_1(e) \cdot \mathsf{in}_2(l_k)$ and at ** we use $\mathsf{out}_1(l_f) = 1$.

Conversely, every clock run r_1 of $\mathcal{A}_1$ on $w_1 \in C_n\Sigma^+$ such that $\mathsf{rwt}(r_1) \neq 0$ must end in l_f, and every clock run r_2 of $\mathcal{A}_2$ on $w_2 \in C_n\Sigma^+$ such that $\mathsf{rwt}(r_2) \neq 0$ must start in some location $l \in \mathcal{L}_2$ such that $\mathsf{in}(l) \neq 0$. Hence, we can uniquely compose a clock run r of $\mathcal{A}$ on w such that $w = w_1; w_2$ and $\mathsf{rwt}(r) = \mathsf{rwt}(r_1) \cdot \mathsf{rwt}(r_2)$. This can be proved in a similar way as previously shown for the other direction.

Using this, we obtain $(\|\mathcal{A}\|_n, w) = \sum_{\substack{w = w_1; w_2 \\ w_1, w_2 \notin \mathcal{E}_n}} (\|\mathcal{A}_1\|_n, w_1) \cdot (\|\mathcal{A}_2\|_n, w_2)$.

However, if we let $w_0 \in \mathcal{E}_n$ and use properness of both $\|\mathcal{A}_1\|_n$ and $\|\mathcal{A}_2\|_n$, we obtain

$$\begin{aligned}
& (\|\mathcal{A}_1\|_n; \|\mathcal{A}_2\|_n, w) \\
= \; & \sum_{w_1; w_2 = w} (\|\mathcal{A}_1\|_n, w_1) \cdot (\|\mathcal{A}_2\|_n, w_2)
\end{aligned}$$

$$
\begin{aligned}
&= \sum_{\substack{w_1;w_2=w \\ w_1,w_2\notin\mathcal{E}_n}} ((\|\mathcal{A}_1\|_n, w_1)\cdot(\|\mathcal{A}_2\|_n, w_2)) \\
&\quad +((\|\mathcal{A}_1\|_n, w_0)\cdot(\|\mathcal{A}_2\|_n, w)) + ((\|\mathcal{A}_1\|_n, w)\cdot(\|\mathcal{A}_2\|_n, w_0)) \\
&= \sum_{\substack{w_1;w_2=w \\ w_1,w_2\notin\mathcal{E}_n}} ((\|\mathcal{A}_1\|_n, w_1)\cdot(\|\mathcal{A}_2\|_n, w_2)) + 0 + 0 \\
&= \sum_{\substack{w_1;w_2=w \\ w_1,w_2\notin\mathcal{E}_n}} (\|\mathcal{A}_1\|_n, w_1)\cdot(\|\mathcal{A}_2\|_n, w_2).
\end{aligned}
$$

Hence, we have $\|\mathcal{A}\|_n = \|\mathcal{A}_1\|_n;\|\mathcal{A}_2\|_n = S_1';S_2'$ and thus $S_1';S_2'$ is $\mathcal{F}$-recognizable.

Next, we show that $s\varepsilon;S_2'$ is $\mathcal{F}$-recognizable. We put $\mathcal{A} = (\mathcal{L}_2, \mathcal{C}, E_2, \mathsf{in}, \mathsf{out}_2, \mathsf{ewt}_2, \mathsf{lwt}_2)$, where $\mathsf{in}(l) = s\cdot\mathsf{in}_2(l)$. Then, we obtain

$$
\begin{aligned}
(\|\mathcal{A}\|_n, w) &= \sum\{\mathsf{rwt}(r) : r \text{ is a clock run of } \mathcal{A} \text{ on } w\} \\
&= \sum\{s\cdot\mathsf{rwt}(r) : r \text{ is a clock run of } \mathcal{A}_2 \text{ on } w\} \\
&= s\cdot\sum\{\mathsf{rwt}(r) : r \text{ is a clock run of } \mathcal{A}_2 \text{ on } w\} \\
&= s\cdot(\|\mathcal{A}_2\|, w) \\
&= s\cdot(S_2', w)
\end{aligned}
$$

by definition of in and distributivity. However, we also have

$$
\begin{aligned}
(s\varepsilon;S_2', w) &= \sum_{w_1;w_2=w} (s\varepsilon, w_1)\cdot(S_2', w_2) \\
&= \sum_{\substack{w_1;w_2=w \\ w_1\in\mathcal{E}_n}} (s\varepsilon, w_1)\cdot(S_2', w_2) + \sum_{\substack{w_1;w_2=w \\ w_1\notin\mathcal{E}_n}} (s\varepsilon, w_1)\cdot(S_2', w_2) \\
&= \sum_{\substack{w_1;w_2=w \\ w_1\in\mathcal{E}_n}} (s\varepsilon, w_1)\cdot(S_2', w_2) + 0 \\
&= \sum_{\substack{w_1;w_2=w \\ w_1\in\mathcal{E}_n}} (s\varepsilon, w_1)\cdot(S_2', w_2) \\
&= s\cdot(S_2', w)
\end{aligned}
$$

due to the fact that $(s\varepsilon, w) = s$ if $w \in \mathcal{E}_n$, and 0 otherwise. For proving $\mathcal{F}$-recognizability of $S_1';s'\varepsilon$, we put $\mathcal{A} = (\mathcal{L}_1, \mathcal{C}, E_1, \mathsf{in}_1, \mathsf{out}, \mathsf{ewt}_1, \mathsf{lwt}_1)$, where $\mathsf{out}(l) = \mathsf{out}_1(l)\cdot s'$. Then, $\|\mathcal{A}\|_n = S_1';s'\varepsilon$, which can be proved as above. Finally, we define $\mathcal{A} = (\mathcal{L}, \mathcal{C}, E, \mathsf{in}, \mathsf{out}, \mathsf{ewt}, \mathsf{lwt})$ by

- $\mathcal{L} = \{l\}$,
- $E = \emptyset$,

- $\mathsf{in}(l) = s$,
- $\mathsf{out}(l) = s'$,
- $\mathsf{lwt}(l) \in \mathcal{F}$.

Clearly, $\|\mathcal{A}\|_n = s\varepsilon; s'\varepsilon$. ■

For showing closure of the class of $\mathcal{F}$-recognizable n-clock series under the Kleene star iteration, we need another normalization method. A weighted timed automaton is said to be *initial-location-normalized* if, for each location l, whenever $\mathsf{in}(l) \neq 0$, then l is a source, i.e., has no ingoing edges.

Lemma 4.11. *For each weighted timed automaton $\mathcal{A}$, there is an initial-location-normalized weighted timed automaton $\mathcal{A}'$ with $(\|\mathcal{A}'\|_n, w) = (\|\mathcal{A}\|_n, w)$ for each $w \in C_n\Sigma^*$.*

PROOF. Let $\mathcal{A} = (\mathcal{L}, \mathcal{C}, E, \mathsf{in}, \mathsf{out}, \mathsf{ewt}, \mathsf{lwt})$ be a weighted timed automaton. For each $l \in \mathcal{L}$, let $\mathrm{cp}(l)$ be the copy of l. Define $\mathcal{A}' = (\mathcal{L}', \mathcal{C}, E', \mathsf{in}', \mathsf{out}', \mathsf{ewt}', \mathsf{lwt}')$, where

- $\mathcal{L}' = \mathcal{L} \cup \{\mathrm{cp}(l) : l \in \mathcal{L} \text{ such that } \mathsf{in}(l) \neq 0\}$,
- $E' = E \cup \{(l, a, \phi, \lambda, l') : \text{ there is some } l_2 \in \mathcal{L} \text{ such that } (l_2, a, \phi, \lambda, l') \in E, l = \mathrm{cp}(l_2)\}$,
- $\mathsf{in}'(l) = \begin{cases} 0 & \text{if } l \in \mathcal{L} \\ \mathsf{in}(l') & \text{if } l = \mathrm{cp}(l') \text{ for some } l' \in \mathcal{L}, \end{cases}$
- $\mathsf{out}'(l) = \begin{cases} \mathsf{out}(l) & \text{if } l \in \mathcal{L} \\ \mathsf{out}(l') & \text{if } l = \mathrm{cp}(l') \text{ for some } l' \in \mathcal{L}, \end{cases}$
- $\mathsf{ewt}'(l, a, \phi, \lambda, l') = \begin{cases} \mathsf{ewt}(l, a, \phi, \lambda, l') & \text{if } (l, a, \phi, \lambda, l') \in E \\ \mathsf{ewt}(l_2, a, \phi, \lambda, l') & \text{if } l = \mathrm{cp}(l_2) \text{ for some } l_2 \in \mathcal{L}, \end{cases}$
- $\mathsf{lwt}'(l) = \begin{cases} \mathsf{lwt}(l) & \text{if } l \in \mathcal{L} \\ \mathsf{lwt}(l') & \text{if } l = \mathrm{cp}(l') \text{ for some } l' \in \mathcal{L}. \end{cases}$

Then $\|\mathcal{A}'\|_n = \|\mathcal{A}\|_n$ is proved by establishing a weight-preserving bijective correspondence between the clock runs of $\mathcal{A}$ and $\mathcal{A}'$. ■

Remark 4.12. Note that unlike classical initial-state-normalizations of weighted finite automata (see e.g. [58]), we do not require a *single* source location l such that $\mathsf{in}(l) = 1$ and $\mathsf{in}(l') = 0$ for each $l' \in \mathcal{L} \backslash \{l\}$. This is due to two reasons. First, we cannot restrict to a *single* source location, as it is not clear how to define the weight function for this location in such a way that it corresponds to the weight functions of all former locations l with $\mathsf{in}(l) \neq 0$. Recall that in the final-location-normalization construction we could use a single sink location, because this has no outgoing edges and hence its location weight does not influence the behaviour of the weighted timed automaton. Thus we are free in choosing any location weight from $\mathcal{F}$ for this sink location. Second, we cannot require the weights for entering the "initial" locations to be equal to 1. This is also due to the location weight functions assigned to locations. In the classical construction by Eilenberg, the weights for entering the initial locations are multiplied with the edge weights, i.e., using a similar construction as in the proof above, we define

$$\mathsf{ewt}(l, a, \phi, \lambda, l') = \begin{cases} \mathsf{ewt}(l, a, \phi, \lambda, l') & \text{if } (l, a, \phi, \lambda, l') \in E \\ \mathsf{in}(l_2) \cdot \mathsf{ewt}(l_2, a, \phi, \lambda, l') & \text{if } l = \mathrm{cp}(l_2) \text{ for some } l_2 \in \mathcal{L}. \end{cases}$$

However, this construction only works for $\mathcal{K}$ being commutative, as in the computation of the running weight of a clock run, between $\mathsf{in}(l)$ and $\mathsf{ewt}(e)$ we have to consider the location weight $\mathsf{lwt}(l)(\delta)$. Another idea is to redefine the weight function of the initial locations l, i.e., multiply the weights $\mathsf{in}(l)$ from the left with the result of $\mathsf{lwt}(l)(\delta)$; but this does not work for e.g. families of linear functions, as the resulting function is no longer in $\mathcal{F}$.

Corollary 4.13. *For each weighted timed automaton $\mathcal{A}$ there is an initial- and final-location-normalized weighted timed automaton $\mathcal{A}_N$ with $(\|\mathcal{A}_N\|_n, w) = (\|\mathcal{A}\|_n, w)$ for each $w \in C_n\Sigma^+$ and $(\|\mathcal{A}_N\|_n, w) = 0$ for each $w \in \mathcal{E}_n$.*

Proof. Follow the constructions of Lemmas 4.9 and 4.11, and note that in $\mathcal{A}_N$ we have $\mathsf{in}(l) = 0$ if and only if $\mathsf{out}(l) \neq 0$ for each location l. ∎

Lemma 4.14. *If $S : C_n\Sigma^* \to K$ is a proper $\mathcal{F}$-recognizable n-clock series, then S^* is $\mathcal{F}$-recognizable.*

Proof. Since S is proper, we have $(S, w) = 0$ if $w \in \mathcal{E}_n$. Thus by Corollary 4.13 there is an initial- and final-location-normalized weighted timed automaton $\mathcal{A} = (\mathcal{L}, \mathcal{C}, E, \mathsf{in}, \mathsf{out}, \mathsf{ewt}, \mathsf{lwt})$ over $\mathcal{K}$, Σ and $\mathcal{F}$ with $\|\mathcal{A}\|_n = S$. We use l_f to denote the sink location of $\mathcal{A}$. We define $\mathcal{A}' = (\mathcal{L}', \mathcal{C}, E', \mathsf{in}', \mathsf{out}', \mathsf{ewt}', \mathsf{lwt}')$, where

- $\mathcal{L}' = \mathcal{L} \dot\cup \{l_\varepsilon\}$,
- $E' = E \cup \{(l, a, \phi, \lambda, l') : l' \in \mathcal{L}, \mathsf{in}(l') \neq 0, (l, a, \phi, \lambda, l_f) \in E\}$,
- $\mathsf{in}'(l) = \begin{cases} \mathsf{in}(l) & \text{if } l \in \mathcal{L} \\ 1 & \text{if } l = l_\varepsilon, \end{cases}$

- $\mathsf{out}'(l) = \begin{cases} \mathsf{out}(l) & \text{if } l \in \mathcal{L} \\ 1 & \text{if } l = l_\varepsilon, \end{cases}$

- $\mathsf{ewt}'(l, a, \phi, \lambda, l') = \begin{cases} \mathsf{ewt}(l, a, \phi, \lambda, l') & \text{if } (l, a, \phi, \lambda, l') \in E \\ \mathsf{ewt}(l, a, \phi, \lambda, l_f) \cdot \mathsf{in}(l') & \text{otherwise}, \end{cases}$

- $\mathsf{lwt}'(l) = \begin{cases} \mathsf{lwt}(l) & \text{if } l \in \mathcal{L} \\ \text{arbitrary but in } \mathcal{F} & \text{otherwise.} \end{cases}$

We prove that $(\|\mathcal{A}'\|_n, w) = ((\|\mathcal{A}\|_n)^*, w)$. First, we let $w \in C_n\Sigma^+$. We provide a weight-preserving bijection between the clock runs of $\mathcal{A}'$ and finite sequences of clock runs of $\mathcal{A}$.

Construction 1 Let $r = (l_0, t_0, \nu_0) \xrightarrow{\delta_1} \xrightarrow{e'_1} \dots \xrightarrow{\delta_k} \xrightarrow{e'_k} (l_k, t_k, \nu_k)$ be a run of $\mathcal{A}'$ on w. Define $\{y_1, ..., y_m\} \subseteq \{1, ..., k\}$ to be the set of index numbers with $l_{y_i} \in \mathcal{L}$ such that $\mathsf{in}(l_{y_i}) \neq 0$ for each $i \in \{1, ..., m\}$. We split the clock run r into $m+1$ clock runs r'_j as follows:

$$r'_1 = (l_0, t_0, \nu_0) \xrightarrow{\delta_1} \xrightarrow{e'_1} \dots \xrightarrow{\delta_{y_1}} \xrightarrow{e'_{y_1}} (l_{y_1}, t_{y_1}, \nu_{y_1})$$

$$r'_j = (l_{y_{j-1}}, t_{y_{j-1}}, \nu_{y_{j-1}}) \xrightarrow{\delta_{y_{j-1}+1}} \xrightarrow{e'_{y_{j-1}+1}} \dots \xrightarrow{\delta_{y_j}} \xrightarrow{e'_{y_j}} (l_{y_j}, t_{y_j}, \nu_{y_j})$$

for each $j \in \{2, ..., m\}$, and

$$r'_{m+1} = (l_{y_m}, t_{y_m}, \nu_{y_m}) \xrightarrow{\delta_{y_m+1}} \xrightarrow{e'_{y_m+1}} \dots \xrightarrow{\delta_k} \xrightarrow{e'_k} (l_k, t_k, \nu_k).$$

Let $j \in \{1, ..., m\}$. Observe that the last edge e'_{y_j} of r'_j is an edge in $E' \setminus E$ since E does not contain any edges to locations l with $\mathsf{in}(l) \neq 0$. Hence, there is an edge $e_{y_j} \in E$ of the form $(l_{y_j-1}, a_{y_j}, \phi_{y_j}, \lambda_{y_j}, l_f)$. We define the clock run r_j from r'_j by replacing the last edge e'_{y_j} by e_{y_j}. We further put $r_{m+1} = r'_{m+1}$. Then, using $\mathsf{ewt}'(e'_{y_j}) = \mathsf{ewt}(e_{y_j}) \cdot \mathsf{in}(l_{y_j})$ and $\mathsf{out}(l_f) = 1$ (*) for each $j \in \{1, ..., m\}$, we obtain

$$\begin{aligned}
&\mathsf{rwt}(r) \\
= \;& \mathsf{in}'(l_0) \cdot \left(\prod_{1 \le i \le k} \mathsf{lwt}'(l_{i-1})(\delta_i) \cdot \mathsf{ewt}'(e'_i) \right) \cdot \mathsf{out}'(l_k) \\
= \;& \mathsf{in}'(l_0) \cdot \left(\prod_{1 \le i \le y_1} \mathsf{lwt}'(l_{i-1})(\delta_i) \cdot \mathsf{ewt}'(e'_i) \right) \cdot \dots \cdot \left(\prod_{y_m < i \le k} \mathsf{lwt}'(l_{i-1})(\delta_i) \cdot \mathsf{ewt}'(e'_i) \right) \cdot \mathsf{out}'(l_k) \\
= \;& \mathsf{in}'(l_0) \cdot \left(\prod_{1 \le i < y_1} \mathsf{lwt}'(l_{i-1})(\delta_i) \cdot \mathsf{ewt}'(e'_i) \right) \cdot \mathsf{lwt}'(l_{y_1-1})(\delta_{y_1}) \cdot \mathsf{ewt}'(e_{y_1}) \cdot \dots
\end{aligned}$$

$$
\begin{aligned}
& \cdot \left(\prod_{y_m < i \leq k} \mathsf{lwt}'(l_{i-1})(\delta_i) \cdot \mathsf{ewt}'(e_i') \right) \cdot \mathsf{out}'(l_k) \\
\overset{\star}{=} \; & \mathsf{in}(l_0) \cdot \left(\prod_{1 \leq i < y_1} \mathsf{lwt}(l_{i-1})(\delta_i) \cdot \mathsf{ewt}(e_i') \right) \cdot \mathsf{lwt}(l_{y_1-1})(\delta_{y_1}) \cdot \mathsf{ewt}(e_{y_1}) \\
& \cdot \mathsf{in}(l_{y-1}) \cdot \ldots \cdot \left(\prod_{y_m < i \leq k} \mathsf{lwt}(l_{i-1})(\delta_i) \cdot \mathsf{ewt}(e_i') \right) \cdot \mathsf{out}(l_k) \\
= \; & \mathsf{rwt}(r_1) \cdot \ldots \cdot \mathsf{rwt}(r_{m+1}).
\end{aligned}
$$

Construction 2 For each $i \in \{1, ..., m\}$, let $r_i = (l_{i,0}, t_{i,0}, \nu_{i,0}) \xrightarrow{\delta_{i,1}} \xrightarrow{e_{i,1}} \ldots \xrightarrow{\delta_{i,k_i}} \xrightarrow{e_{i,k_i}} (l_f, t_{i,k_i}, \nu_{i,k_i})$ be a clock run of $\mathcal{A}$ on w_i such that $w = w_1; ...; w_m$. For each $i \in \{1, ..., m-1\}$, we let r_i' be the clock run that is obtained from r_i by replacing the last edge e_{i,k_i} by an edge e_{i,k_i}' of the form $(l_{i,k_i-1}, a_{i,k_i}, \phi_{i,k_i}, \lambda_{i,k_i}, l_{i+1,0})$. Notice that such an edge exists due to the definition of E'. By definition, we have $\mathsf{ewt}'(e_{i,k_i}') = \mathsf{ewt}(e_{i,k_i})$. The assumption $w = w_1; ...; w_m$ implies $(t_{i+1,0}, \nu_{i+1,0}) = (t_{i,k_i}, \nu_{i,k_i})$ for each $i \in \{1, ..., m-1\}$. Thus we can compose the clock runs $r_1', ..., r_{m-1}', r_m$, obtaining a new run $r = r_1' ... r_{m-1}' r_m$, which is a clock run of $\mathcal{A}'$ on w. Using the same lines of argumentation as above, we can prove $\mathsf{rwt}(r) = \mathsf{rwt}(r_1) \cdot ... \cdot \mathsf{rwt}(r_m)$.

Note that construction 1 and 2 are inverse to each other. Altogether, we have

$$
\begin{aligned}
& (\|\mathcal{A}'\|_n, w) \\
= \; & \sum \{\mathsf{rwt}(r) : r \text{ is a clock run of } \mathcal{A}' \text{ on } w\} \\
\overset{\star}{=} \; & \sum \{\mathsf{rwt}(r_1) \cdot \ldots \cdot \mathsf{rwt}(r_m) : r_i \text{ is a clock run of } \mathcal{A} \text{ on } w_i, \\
& \text{for each } i = 1, ..., m, \text{ such that } w = w_1; ...; w_m, 1 \leq m \leq |w|\} \\
= \; & \sum_{1 \leq m \leq |w|} \; \sum_{w = w_1; ...; w_m} \sum \{\mathsf{rwt}(r_1) \cdot \ldots \cdot \mathsf{rwt}(r_m) : r_i \text{ is a clock run} \\
& \text{of } \mathcal{A} \text{ on } w_i \text{ for each } i = 1, ..., m\} \\
= \; & \sum_{1 \leq m \leq |w|} \; \sum_{w = w_1; ...; w_m} \sum \{\mathsf{rwt}(r_1) : r_1 \text{ is a clock run of } \mathcal{A} \text{ on } w_1\} \\
& \cdot \ldots \cdot \sum \{\mathsf{rwt}(r_m) : r_m \text{ is a clock run of } \mathcal{A} \text{ on } w_m\} \\
= \; & \sum_{1 \leq m \leq |w|} \; \sum_{w = w_1; ...; w_m} (\|\mathcal{A}\|_n, w_1) \cdot \ldots \cdot (\|\mathcal{A}\|_n, w_m) \\
\overset{\star\star}{=} \; & \sum_{1 \leq m \leq |w|} ((\|\mathcal{A}\|_n)^m, w) \\
= \; & ((\|\mathcal{A}\|_n)^*, w)
\end{aligned}
$$

where * uses the weight-preserving bijective correspondence described above, and ** is an application of Lemma 4.4.

Now let $w \in \mathcal{E}_n$. In $\mathcal{A}'$ there is exactly one location l such that $\mathsf{in}'(l) \neq 0$ and $\mathsf{out}'(l) \neq 0$, namely l_ε. In fact, we have $\mathsf{in}'(l_\varepsilon) = \mathsf{out}'(l_\varepsilon) = 1$, and thus we have $(\|\mathcal{A}'\|_n, w) = 1 = ((\|\mathcal{A}\|_n)^*, w)$ Hence, $S^* = \|\mathcal{A}'\|_n$ is $\mathcal{F}$-recognizable. ■

Theorem 4.15. *Let $S : C_n\Sigma^* \to K$ be an $\mathcal{F}$-rational n-clock series. Then S is $\mathcal{F}$-recognizable.*

PROOF. Follows from Lemmas 4.6, 4.7, 4.8, 4.10 and 4.14. ■

4.4 From Recognizability to Rationality

In this section, we show the other inclusion of the Kleene-Schützenberger theorem, namely that each $\mathcal{F}$-recognizable n-clock series is $\mathcal{F}$-rational. We do this by solving a system of equations induced by the given weighted timed automaton. The solution of the system corresponds to an $\mathcal{F}$-rational n-clock series. Before we present the actual result, we give some lemmas. Let $\mathcal{A} = (\mathcal{L}, \mathcal{C}, E, \mathsf{in}, \mathsf{out}, \mathsf{ewt}, \mathsf{lwt})$ be a weighted timed automaton over $\mathcal{K}$, Σ and $\mathcal{F}$. For any two locations $l, l' \in \mathcal{L}$, we define $\mathcal{A}_{l,l'} = (\mathcal{L}, \mathcal{C}, E, \mathsf{in}', \mathsf{out}', \mathsf{ewt}, \mathsf{lwt})$, where

$$\mathsf{in}'(l_1) = \begin{cases} 1 & \text{if } l_1 = l \\ 0 & \text{otherwise} \end{cases}$$

and

$$\mathsf{out}'(l_1) = \begin{cases} 1 & \text{if } l_1 = l' \\ 0 & \text{otherwise.} \end{cases}$$

Lemma 4.16. *Let $\mathcal{A} = (\mathcal{L}, \mathcal{C}, E, \mathsf{in}, \mathsf{out}, \mathsf{ewt}, \mathsf{lwt})$ be an initial- and final-location-normalized weighted timed automaton. If l_f denotes the sink location of $\mathcal{A}$, then we have*

$$\|\mathcal{A}\|_n = \sum_{\substack{l \in \mathcal{L} \\ \mathsf{in}(l) \neq 0}} (\mathsf{in}(l))\varepsilon; \|\mathcal{A}_{l,l_f}\|_n.$$

PROOF. For $w \in \mathcal{E}_n$, the assertion is clear. So let $w \in C_n\Sigma^+$. Then we have

$$\begin{aligned} (\|\mathcal{A}\|_n, w) &= \sum \{\mathsf{rwt}(r) : r \text{ is a clock run of } \mathcal{A} \text{ on } w\} \\ &= \sum \{\mathsf{rwt}(r) : r \text{ is a clock run of } \mathcal{A} \text{ on } w \text{ from } l \text{ to } l_f \text{ for some } l \in \mathcal{L} \text{ such} \\ &\qquad \text{that } \mathsf{in}(l) \neq 0\} \end{aligned}$$

$$\begin{aligned}
&= \sum_{\substack{l\in\mathcal{L}\\ \mathsf{in}(l)\neq 0}} \sum\{\mathsf{rwt}(r) : r \text{ is a clock run of } \mathcal{A} \text{ on } w \text{ from } l \text{ to } l_f\}\\
&= \sum_{\substack{l\in\mathcal{L}\\ \mathsf{in}(l)\neq 0}} \sum\{\mathsf{in}(l)\cdot\mathsf{rwt}(r) : r \text{ is a clock run of } \mathcal{A}_{l,l_f} \text{ on } w\}\\
&= \sum_{\substack{l\in\mathcal{L}\\ \mathsf{in}(l)\neq 0}} \mathsf{in}(l)\cdot\sum\{\mathsf{rwt}(r) : r \text{ is a clock run of } \mathcal{A}_{l,l_f} \text{ on } w\}\\
&= \sum_{\substack{l\in\mathcal{L}\\ \mathsf{in}(l)\neq 0}} \mathsf{in}(l)\cdot(\|\mathcal{A}_{l,l_f}\|_n, w)\\
&\stackrel{\star}{=} \sum_{\substack{l\in\mathcal{L}\\ \mathsf{in}(l)\neq 0}} ((\mathsf{in}(l))\varepsilon; \|\mathcal{A}_{l,l_f}\|_n, w)\\
&= \left(\sum_{\substack{l\in\mathcal{L}\\ \mathsf{in}(l)\neq 0}} (\mathsf{in}(l))\varepsilon; \|\mathcal{A}_{l,l_f}\|_n, w\right),
\end{aligned}$$

where at * we use the fact that for each n-clock series $S : C_n\Sigma^* \to K$ and $k \in K$, we have $k \cdot S = k\varepsilon; S$. ∎

Lemma 4.17. *Let $\mathcal{A} = (\mathcal{L}, \mathcal{C}, E, \mathsf{in}, \mathsf{out}, \mathsf{ewt}, \mathsf{lwt})$ be a weighted timed automaton and let $l_f \in \mathcal{L}$. Then, for every $l \in \mathcal{L}$ we have*

$$\|\mathcal{A}_{l,l_f}\|_n = \begin{cases} \sum\limits_{(l,a,\phi,\lambda,l')\in E} \|\mathcal{A}_{\langle \mathsf{lwt}(l),k,a,\phi,\lambda\rangle}\|_n; \|\mathcal{A}_{l',l_f}\|_n + 1\varepsilon & \text{if } l = l_f,\\ \sum\limits_{(l,a,\phi,\lambda,l')\in E} \|\mathcal{A}_{\langle \mathsf{lwt}(l),k,a,\phi,\lambda\rangle}\|_n; \|\mathcal{A}_{l',l_f}\|_n & \text{otherwise,} \end{cases}$$

where $k = \mathsf{ewt}(l, a, \phi, \lambda, l')$.

PROOF. First we assume $l \neq l_f$. For $w \in \mathcal{E}_n$, the assertion is obvious. So let $w = (t_0, \nu_0)(a_1, t_1, \nu_1)...(a_m, t_m, \nu_m) \in C_n\Sigma^+$. We show that there is a weight-preserving bijective correspondence between the set of clock runs of $\mathcal{A}_{l,l_f}$ on w and the set of pairs of clock runs of $\mathcal{A}_{\langle \mathsf{lwt}(l),k,a_1,\phi,\lambda\rangle}$ on $w_1 = (t_0, \nu_0)(a_1, t_1, \nu_1)$ and clock runs of $\mathcal{A}_{l',l_f}$ on $w_2 = (t_1, \nu_1)(a_2, t_2, \nu_2)...(a_m, t_m, \nu_m)$ for some $(l, a_1, \phi, \lambda, l') \in E$. Observe that $w = w_1; w_2$.

Let $r = (l_0, t_0, \nu_0) \xrightarrow{\delta_1}\xrightarrow{e_1} (l_1, t_1, \nu_1) \xrightarrow{\delta_2}\xrightarrow{e_2} \dots \xrightarrow{\delta_m}\xrightarrow{e_m} (l_m, t_m, \nu_m)$ be a clock run of $\mathcal{A}_{l,l_f}$ on w, where $l_0 = l$ and $l_m = l_f$. We split r after the first transition, obtaining the clock run $r_1 = (l_0, t_0, \nu_0) \xrightarrow{\delta_1}\xrightarrow{e_1} (l_1, t_1, \nu_1)$ on w_1 and the clock run $r_2 = (l_1, t_1, \nu_1) \xrightarrow{\delta_2}\xrightarrow{e_2} \dots \xrightarrow{\delta_m}\xrightarrow{e_m} (l_m, t_m, \nu_m)$ on w_2. Assuming $e_1 = (l_0, a_1, \phi_1, \lambda_1, l_1)$ and $k = \mathsf{ewt}(e_1)$, we know that r_1 is also a clock run of $\mathcal{A}_{\langle \mathsf{lwt}(l),k,a_1,\phi_1,\lambda_1\rangle}$ on w_1 with

$\mathsf{rwt}(r_1) = 1 \cdot \mathsf{lwt}(l_0)(\delta_1) \cdot \mathsf{ewt}(e_1) \cdot 1$. In addition, r_2 is a clock run of $\mathcal{A}_{l_1,l_m}$ on w_2 with $\mathsf{rwt}(r_2) = 1 \cdot \prod_{2\leq i\leq m} \mathsf{lwt}(l_{i-1})(\delta_i) \cdot \mathsf{ewt}(e_i) \cdot 1$. Then, we obtain

$$\begin{aligned}
& \mathsf{rwt}(r_1) \cdot \mathsf{rwt}(r_2) \\
= \; & 1 \cdot \mathsf{lwt}(l_0)(\delta_1) \cdot \mathsf{ewt}(e_1) \cdot 1 \cdot 1 \cdot \prod_{2\leq i\leq m} \mathsf{ewt}_{l_{i-1}}(\delta_i) \cdot \mathsf{ewt}(e_i) \cdot 1 \\
= \; & 1 \cdot \prod_{1\leq i\leq m} \mathsf{lwt}(l_{i-1})(\delta_i) \cdot \mathsf{ewt}(e_i) \cdot 1 \\
= \; & \mathsf{rwt}(r).
\end{aligned}$$

Clearly, this construction establishes a weight-preserving bijective correspondence as claimed. Also notice that r_1 is the only clock run of $\mathcal{A}_{\langle \mathsf{lwt}(l),k,a_1,\phi,\lambda_1\rangle}$ on w_1 with a running weight different from 0. Thus we have $(\|\mathcal{A}_{\langle \mathsf{lwt}(l),k,a_1,\phi,\lambda_1\rangle}\|_n, w_1) = \mathsf{rwt}(r_1)$. Using this, associativity (*) and distributivity (**), we obtain

$$\begin{aligned}
& (\|\mathcal{A}_{l,l_f}\|_n, w) \\
= \; & \sum\{\mathsf{rwt}(r) \mid r \text{ is a clock run of } \mathcal{A}_{l,l_f} \text{ from } l \text{ to } l_f \text{ on } \} \\
\overset{\star}{=} \; & \sum_{(l,a,\phi,\lambda,l')\in E} \sum_{w_1;w_2=w} \sum\{(\|\mathcal{A}_{\langle \mathsf{lwt}(l),k,a,\phi,\lambda\rangle}\|_n, w_1) \cdot \mathsf{rwt}(r_2) : r_2 \text{ is a clock run of } \mathcal{A}_{l',l_f} \\
& \qquad \text{on } w_2\} \\
\overset{\star\star}{=} \; & \sum_{(l,a,\phi,\lambda,l')\in E} \sum_{w_1;w_2=w} (\|\mathcal{A}_{\langle \mathsf{lwt}(l),k,a,\phi,\lambda\rangle}\|_n, w_1) \cdot \sum\{\mathsf{rwt}(r_2) : r_2 \text{ is a clock run of } \mathcal{A}_{l',l_f} \\
& \qquad \text{on } w_2\} \\
= \; & \sum_{(l,a,\phi,\lambda,l')\in E} \sum_{w_1;w_2=w} (\|\mathcal{A}_{\langle \mathsf{lwt}(l),k,a,\phi,\lambda\rangle}\|_n, w_1) \cdot (\|\mathcal{A}_{l',l_f}\|_n, w_2) \\
= \; & \sum_{(l,a,\phi,\lambda,l')\in E} (\|\mathcal{A}_{\langle \mathsf{lwt}(l),k,a,\phi,\lambda\rangle}\|_n ; \|\mathcal{A}_{l',l_f}\|_n, w)
\end{aligned}$$

where $k = \mathsf{ewt}(l, a, \phi, \phi, l')$.

Now assume $l = l_f$. Let $w \in \mathcal{E}_n$. Then we have

$$\begin{aligned}
(\|\mathcal{A}_{l,l_f}\|_n, w) &= \mathsf{in}(l) \cdot \mathsf{out}(l_f) \\
&= (1\varepsilon, w) \\
&= 0 + (1\varepsilon, w) \\
&= (\sum_{(l,a,\phi,\lambda,l')\in E} \|\mathcal{A}_{\langle \mathsf{lwt}(l),k,a,\phi,\lambda\rangle}\|_n ; \|\mathcal{A}_{l',l_f}\|_n, w) + (1\varepsilon, w) \\
&= (\sum_{(l,a,\phi,\lambda,l')\in E} \|\mathcal{A}_{\langle \mathsf{lwt}(l),k,a,\phi,\lambda\rangle}\|_n ; \|\mathcal{A}_{l',l_f}\|_n + 1\varepsilon, w).
\end{aligned}$$

For $w \in C_n\Sigma^+$, we adopt the proof from above and use $(1\varepsilon, w) = 0$. ∎

The objective of these lemmas is to provide the basis for building a system of linear equations that represents the behaviour of a given weighted timed automaton. The solution of this system corresponds to an $\mathcal{F}$-rational clock series that is equivalent to the behaviour of the weighted timed automaton. However, we need to show that it is guaranteed that there is such a solution. Lemma 4.3 supplies us with an even stronger result, namely that there is a *unique* solution. Finally, we present the crucial property between $\mathcal{F}$-recognizable and $\mathcal{F}$-rational n-clock series. For proving it, we use Lemmas 4.16, 4.17 and 4.3.

Theorem 4.18. *If $S : C_n\Sigma^* \to K$ is an $\mathcal{F}$-recognizable n-clock series, then S is $\mathcal{F}$-rational.*

PROOF. Let $S' : C_n\Sigma^* \to K$ be $\mathcal{F}$-recognizable and write $S' = S + k\varepsilon$, where S is proper. Let $\mathcal{A}$ be an initial- and final-location-normalized weighted timed automaton such that $\|\mathcal{A}\|_n = S$. We use l_f to denote the sink location of $\mathcal{A}$. By Lemma 4.16, it suffices to show that the clock series $\|\mathcal{A}_{l,l_f}\|_n$ is $\mathcal{F}$-rational for any $l \in \mathcal{L}$ such that $\mathsf{in}(l) \neq 0$. Using Lemma 4.17, we build a system of linear equations as follows. We first note that $\|\mathcal{A}_{l_f,l_f}\| = 1\varepsilon$ as there are no outgoing edges from l_f. For every $l \in \mathcal{L}\backslash\{l_f\}$ we consider the equation

$$\|\mathcal{A}_{l,l_f}\|_n = \sum_{(l,a,\phi,\lambda,l')\in E} \|\mathcal{A}_{\langle \mathsf{lwt}(l),k,a,\phi,\lambda\rangle}\|_n ; \|\mathcal{A}_{l',l_f}\|_n,$$

where $k = \mathsf{ewt}(l, a, \phi, \lambda, l')$. In each of these equations, we replace the occurrence of $\|\mathcal{A}_{l_f,l_f}\|$ by 1ε. In this system of linear equations of size $|\mathcal{L}|$:

- clock series of the form $\|\mathcal{A}_{l,l_f}\|_n$ correspond to unknown variables, and
- the clock series $\|\mathcal{A}_{\langle \mathsf{lwt}(l),k,a,\phi,\lambda\rangle}\|_n$ correspond to the coefficients of the system, as does the clock series 1ε. By Lemma 4.7 and definition of $\mathcal{F}$-rational clock series, these clock series are $\mathcal{F}$-rational.

We show that clock series of the form $\|\mathcal{A}_{l,l_f}\|_n$ are also $\mathcal{F}$-rational. Formally, this corresponds to solving the system of linear equations. We let $l_1 < l_2 < ... < l_m$ be an arbitrary order on $\mathcal{L}\backslash\{l_f\}$. We solve the equation $\|\mathcal{A}_{l_m,l_f}\|$ (with $\|\mathcal{A}_{l_m,l_f}\|$ as possible unknown variable) as follows. We split the sum over the edges as follows

$$\begin{aligned} \|\mathcal{A}_{l_m,l_f}\|_n = & \sum_{(l_m,a,\phi,\lambda,l_m)\in E} \|\mathcal{A}_{\langle \mathsf{lwt}(l_m),k,a,\phi,\lambda\rangle}\|_n ; \|\mathcal{A}_{l_m,l_f}\|_n \\ & + \sum_{\substack{(l_m,a',\phi',\lambda',l')\in E \\ l_m \neq l'}} \|\mathcal{A}_{\langle \mathsf{lwt}(l),k',a',\phi',\lambda'\rangle}\|_n ; \|\mathcal{A}_{l',l_f}\|_n \end{aligned}$$

where $k = \mathsf{ewt}(l_m, a, \phi, \lambda, l_m)$, $k' = \mathsf{ewt}(l_m, a', \phi', \lambda', l')$. Recall that each series of the form $\|\mathcal{A}_{\langle \mathsf{lwt}(l_m),k,a,\phi,\lambda\rangle}\|_n$ corresponds to the $\mathcal{F}$-monomial $\langle \mathsf{lwt}(l_m), k, a, \phi, \lambda\rangle$ and hence is proper. Thus, if the first sum is not empty, we may apply Lemma 4.3 and obtain

$$\|\mathcal{A}_{l_m,l_f}\|_n = \Big(\sum_{(l_m,a,\phi,\lambda,l_m)\in E} \|\mathcal{A}_{\langle \mathsf{lwt}(l_m),k,a,\phi,\lambda\rangle}\|_n \Big)^* \\ ; \sum_{\substack{(l_m,a',\phi',\lambda',l')\in E \\ l_m\neq l'}} \|\mathcal{A}_{\langle \mathsf{lwt}(l_m),k',a',\phi',\lambda'\rangle}\|_n ; \|\mathcal{A}_{l',l_f}\|_n.$$

If, on the other hand, the first sum is empty, i.e., equal to 0ε, the clock series $\|\mathcal{A}_{l_m,l_f}\|_n$ disappears from the right hand side of the equation. Hence, in both cases, in the right hand side of the equation there are only unknown variables $\|\mathcal{A}_{l',l_f}\|$ with $l' \neq l_m$. We now substitute $\|\mathcal{A}_{l_m,l_f}\|_n$ in the other $m-1$ equations. We repeat the procedure of solving the equations step by step and observe that, indeed, at each step the conditions of Lemma 4.3 are satisfied. In the last step, we obtain a clock series $\|\mathcal{A}_{l_1,l_f}\|_n$ that can be expressed using monomials, $\mathcal{F}$-monomials and the rational operations. Thus, altogether, we have shown that $\|\mathcal{A}\| = S$ is $\mathcal{F}$-rational. However, for each $k \in K$, the monomial $k\varepsilon$ is also $\mathcal{F}$-rational by definition, and thus also is S'. ∎

We are finally ready to present a Kleene-Schützenberger theorem for the class of weighted timed automata.

Theorem 4.19. *Let $S : C_n\Sigma^* \to K$ be an n-clock series. Then S is $\mathcal{F}$-recognizable if and only if S is $\mathcal{F}$-rational.*

PROOF. This follows immediately from Theorems 4.15 and 4.18. ∎

4.5 From Clock Series to Timed Series

As mentioned before, the clock semantics is used for defining the concatenation operation in a natural way. However, research in the real-time community focuses on *timed languages* rather than clock languages. In this section, we show that a Kleene-Schützenberger theorem can also be given for timed series.

The use of timed semantics sacrifices some significant information concerning the values of the clock variables, which precludes us from defining the notion of rationality for timed series in the same way as for clock series. For this reason, we use the approach of Bouyer and Petit [29], and introduce a projection that maps $\mathcal{F}$-recognizable clock series to $\mathcal{F}$-recognizable timed series. Let $\pi : C_n\Sigma^* \to T\Sigma^*$ be the partial function defined by $\pi\big((t_0,\nu_0)(a_1,t_1,\nu_1)...(a_k,t_k,\nu_k)\big) = (a_1,t_1)...(a_k,t_k)$ if $(t_0,\nu_0) = (0,0^n)$ for each $(t_0,\nu_0)(a_1,t_1,\nu_1)...(a_k,t_k,\nu_k) \in C_n\Sigma^*$, undefined otherwise. We extend π to a function $\bar{\pi} : \mathcal{K}^{\mathcal{F}-rec}\langle\langle C_n\Sigma^*\rangle\rangle \to \mathcal{K}^{\mathcal{F}-rec}\langle\langle T\Sigma^*\rangle\rangle : S \mapsto \bar{\pi}(S)$, where

$$(\bar{\pi}(S), v) = \sum_{\substack{w\in C_n\Sigma^* \\ \pi(w)=v}} (S, w)$$

for each timed word $v \in T\Sigma^*$. Notice that the sum in the equation is finite: for each timed word v there is only a finite number of runs of a weighted timed automaton on v. Each of these runs uniquely determines an n-clock word w such that $\pi(w) = v$. For each other n-clock word w with $\pi(w) = v$ there is no such run and thus $(S, w) = 0$. In fact, we need to restrict $\bar{\pi}$ to *$\mathcal{F}$-recognizable* n-clock series, because for arbitrary n-clock series this property does not hold.

A timed series $\mathcal{T} : T\Sigma^* \to K$ is *$\mathcal{F}$-rational over $\mathcal{K}$ and Σ* if it can be defined by a single application of $\bar{\pi}$ to an $\mathcal{F}$-rational n-clock series $S : C_n\Sigma^* \to K$ over $\mathcal{K}$ and Σ, i.e., $\mathcal{T} = \bar{\pi}(S)$. The following lemma gives the relation between $\mathcal{F}$-recognizable timed series and $\mathcal{F}$-recognizable n-clock series.

Lemma 4.20. *For each weighted timed automaton $\mathcal{A}$, we have $(\|\mathcal{A}\|, v) = (\bar{\pi}(\|\mathcal{A}\|_n), v)$ for every timed word $v \in T\Sigma^*$.*

PROOF. Let $\mathcal{A}$ be a weighted timed automaton over $\mathcal{K}$, Σ and $\mathcal{F}$ and $v = (a_1, t_1)...(a_k, t_k) \in T\Sigma^*$ be a timed word. Let $r = (l_0, \nu_0) \xrightarrow{\delta_1} \xrightarrow{e_1} ... \xrightarrow{\delta_k} \xrightarrow{e_k} (l_k, \nu_k)$ be a run of $\mathcal{A}$ on v. We define the clock run r' by $(l_0, \tau_0, \nu_0) \xrightarrow{\delta_1} \xrightarrow{e_1} ... \xrightarrow{\delta_k} \xrightarrow{e_k} (l_k, \tau_k, \nu_k)$, where $\tau_0 = 0$ and $\tau_i = \tau_{i-1} + \delta_i$ for each $i \in \{1, ..., k\}$. Hence, $\tau_i = t_i$ for each $i \in \{1, ..., k\}$ and r' is a clock run of $\mathcal{A}$ on $(t_0, \nu_0)(a_1, t_1, \nu_1)...(a_k, t_k, \nu_k) \in C_n\Sigma^*$. Clearly, $\pi((t_0, \nu_0)(a_1, t_1, \nu_1)...(a_k, t_k, \nu_k)) = v$. Moreover, we have $\mathsf{rwt}(r') = \mathsf{rwt}(r)$.

Now, let $w = (t_0, \nu_0)(a_1, t_1, \nu_1)...(a_k, t_k, \nu_k) \in C_n\Sigma^*$ be an n-clock word such that $t_0 = 0$ and $\nu_0 = 0^{\mathcal{C}}$ and $r = (l_0, t_0, \nu_0) \xrightarrow{\delta_1} \xrightarrow{e_1} ... \xrightarrow{\delta_k} \xrightarrow{e_k} (l_k, t_k, \nu_k)$ be a clock run of $\mathcal{A}$ on w. We let r' be the run that is obtained by removing the second element from each state in r, i.e., r' is of the form $(l_0, \nu_0) \xrightarrow{\delta_1} \xrightarrow{e_1} ... \xrightarrow{\delta_k} \xrightarrow{e_k} (l_k, \nu_k)$. Clearly, r' is a run of $\mathcal{A}$ on $(a_1, t_1)...(a_k, t_k) \in T\Sigma^*$ and we have $\pi(w) = (a_1, t_1)...(a_k, t_k)$. Furthermore, $\mathsf{rwt}(r') = \mathsf{rwt}(r)$.

Altogether, we have shown that for any clock run of $\mathcal{A}$ on an n-clock word w, there is a run of $\mathcal{A}$ on v such that $\pi(w) = v$ and vice versa. One can easily see that applying both constructions back and forth to a run r, we obtain r again. Moreover, both runs have the same running weight. Hence, the correspondence between the two kinds of runs is bijective and weight-preserving. In the following, we use this at $\star$.

$$
\begin{aligned}
(\|\mathcal{A}\|, v) &= \sum \{\mathsf{rwt}(r) : r \text{ is a run of } \mathcal{A} \text{ on } v\} \\
&\stackrel{\star}{=} \sum \{\mathsf{rwt}(r) : r \text{ is a clock run of } \mathcal{A} \text{ on } w \text{ such that } \pi(w) = v\} \\
&= \sum_{\pi(w)=v} \sum \{\mathsf{rwt}(r) : r \text{ is a clock run of } \mathcal{A} \text{ on } w\} \\
&= \sum_{\pi(w)=v} (\|\mathcal{A}\|_n, w) \\
&= (\bar{\pi}(\|\mathcal{A}\|_n), v).
\end{aligned}
$$

■

So, this implies that also $\mathcal{F}$-recognizable timed series correspond to a single application of $\bar{\pi}$ to an $\mathcal{F}$-recognizable n-clock series. As a result, we obtain a Kleene-Schützenberger theorem also for the usual timed semantics of weighted timed automata.

Theorem 4.21. *Let $\mathcal{T} : T\Sigma^* \to K$ be a timed series. Then $\mathcal{T}$ is $\mathcal{F}$-recognizable if and only if $\mathcal{T}$ is $\mathcal{F}$-rational.*

PROOF. The definition of $\mathcal{F}$-rational timed series and Lemma 4.20 ensure that both $\mathcal{F}$-rational and $\mathcal{F}$-recognizable timed series correspond to a single application of $\bar{\pi}$ to an $\mathcal{F}$-rational ($\mathcal{F}$-recognizable, respectively) n-clock series. This and Theorem 4.19 imply the result. ■

4.6 Conclusion

In this chapter, we presented a Kleene-Schützenberger theorem for weighted timed automata, which can be seen as a timed and weighted analogue to the famous Kleene theorem for finite automata over words [76]. Not only is the Kleene theorem one of the most significant theorems in formal language theory. Rational timed series may also be used as a convenient formalism for specifying the behaviour of weighted timed automata. In this regard, we point out that the translation procedures from rational timed series to weighted timed automata and vice versa are effective. In future work one may investigate the complexity bounds of our constructions. Also, we would like to remark that the Kleene-Schützenberger theorem we presented here and all the constructions are close to the classical untimed case, albeit different in some important aspects. The definition of the concatenation operation (adopted from Bouyer and Petit [29]) is simple and similar to the classical concatenation operation. The formalism we presented is an intuitive algebraic characterization of recognizable timed series. This argues for the clock semantics we used here. However, we do not want to conceil that one may also argue against the clock semantics for the following reason. In the theory of formal languages, regular expressions do not disclose any internal description, whereas here, by using the clock semantics, we mention explicitly some internal information in form of the current values of the clock variables. Hence, it may be of interest to use other approaches that provide a Kleene theorem for the class of timed automata, e.g. the latest approach of Asarin and Dima [14], which does not make use of information external to the timed language.

Another direction for future work on this field concerns the model of weighted timed automata with infinite behaviour, which is of great interest particularly in the area of verification. We note that Bouyer and Petit [29] introduced the clock semantics approach to obtain a Kleene-type theorem for timed automata over *infinite* timed words. However, in the weighted setting, one has to compute the running weight of infinite runs, which may lead to problems of convergence. Possibly, one may solve this using *complete* semirings [59, 60] or distributive lattices [52], which allow for infinite sums and products, or considering models with a discounting factor [42, 51, 56].

5 A Büchi Theorem for Weighted Timed Automata

5.1 Introduction

In this chapter, we will define a logic for the specification of timed series. Then we aim to to show that this logic is expressively equivalent to weighted timed automata. The first result of this kind - the equivalence between finite automata and sentences in MSO logic over some alphabet Σ, denoted by $\mathsf{MSO}(\Sigma)$, was obtained by Büchi [36]. This logical characterization of recognizable languages is one of the most fundamental theorems in theoretical computer science. It is also of great practical interest: a specification expressed by a MSO formula often is much easier to read and understand than an automaton. By Büchi's theorem, each MSO formula corresponds to an automaton. The most important questions that arise in the context of specification are the satisfiability of a formula, and the model checking problem, i.e., the question whether all behaviours of an automaton satisfy the formula. Now, owing to Büchi's theorem, these questions can be answered using well known methods from automata theory.

For the class of WFA-recognizable series, a Büchi theorem has been presented by Droste and Gastin [46, 49]. They introduce a weighted MSO logic for characterizing the behaviours of weighted finite automata defined over a semiring. They extend classical MSO logic with formulas of the form k (where k is an element of the semiring), which may be used to define the weight of a transition of a weighted finite automaton. They show that weighted finite automata are expressively equivalent to a certain fragment of this logic. Recently, this result has been generalized to weighted settings of infinite words [54, 55], trees [57], pictures [85], traces [86], texts [83] and nested words [84].

Here, we aim to generalize the result to timed series. The basis of our work is a Büchi theorem for the class of timed automata by Wilke [107]. For this result, Wilke introduces a timed extension of classical MSO logic. The intuitive idea is to extend $\mathsf{MSO}(\Sigma)$ with formulas of the form $d(y, z) \sim c$, called *distance predicates*, where y, z are first-order variables, $\sim \in \{<, \leq, =, \geq, >\}$ and $c \in \mathbb{N}$. A formula of this form, interpreted over timed words, is supposed to express that the time distance between the positions y and z satisfies the constraint $\sim c$. However, it is shown by Alur and Henzinger [95], that the unrestricted use of distance predicates leads to an undecidable theory. Moreover, since TA-recognizable timed languages are not closed under complement, one cannot expect to find a full MSO logic that is expressively complete for timed automata [107]. For this reason, Wilke restricts the use of distance predicates. He introduced *relative distance predicates* of the form $\overleftarrow{\mathsf{d}}(D, y) \sim c$, where D is a second-order variable, which may only be existentially quantified. Furthermore, this may only be done at the beginning of a formula. The resulting logic is known as *relative distance logic*. Wilke [107] shows that timed languages definable in this logic can be fully characterized in terms of timed automata.

In the next section, we recall the definition of the relative distance logic. For technical simplicity, we do this in two steps. First, we define an auxiliary logic, which is an extension of $\mathsf{MSO}(\Sigma)$ with relative distance predicates. However, within this logic, the second-order variables used as first argument in a relative distance predicate are interpreted as constants. The result is a full MSO logic denoted by $\mathsf{MSO}(T\Sigma^*)$. Second, we define the relative distance logic $\mathcal{L}\overleftarrow{\mathsf{d}}(\Sigma)$, where the second-order variables used as argument in relative distance predicates may be existentially quantified at the beginning of the formula. Then, we will turn to the definition of the weighted relative distance logic. We extend $\mathcal{L}\overleftarrow{\mathsf{d}}(\Sigma)$ with two kinds of weighted formulas of the form k (where $k \in K$) and $f(y)$ (where $f \in \mathcal{F}$ and y is a first-order variable), the semantics of which correspond to the weights of edges and locations, respectively, in weighted timed automata. For proving a Büchi theorem, we stepwisely define a fragment of our logic and show that the semantics of sentences definable in this fragment precisely correspond to the behaviours of weighted timed automata.

5.2 Weighted Relative Distance Logic

We recall the syntax and semantics of the relative distance logic over Σ. As mentioned in the introduction, we do this in two steps. We start with the definition of the **underlying auxiliary logic** $\mathsf{MSO}(T\Sigma^*)$. Formulas of $\mathsf{MSO}(T\Sigma^*)$ are defined by the following grammar

$$\varphi ::= P_a(y) \mid y = z \mid y < z \mid y \in X \mid \overleftarrow{\mathsf{d}}(D, y) \sim c \mid \neg\varphi \mid \varphi \vee \varphi \mid \exists y.\varphi \mid \exists X.\varphi,$$

where y, z are first-order variables, X, D are second-order variables, $a \in \Sigma$, $c \in \mathbb{N}$ and $\sim \in \{<, \leq, =, \geq, >\}$. Formulas of the form $\overleftarrow{\mathsf{d}}(D, y) \sim c$ are called *relative distance predicates*. Notice that the syntax of $\mathsf{MSO}(T\Sigma^*)$ does not allow for the quantification of the first argument D of a relative distance predicate. For this reason, we may temporarily interprete D as a constant and note that $\mathsf{MSO}(T\Sigma^*)$ is a full MSO logic. As usual, we may use true, $\varphi \wedge \psi$, $\varphi \longrightarrow \psi$, $\varphi \longleftrightarrow \psi$, $\forall y.\varphi$ and $\forall X.\varphi$ as abbreviations for $\neg\varphi \vee \varphi$, $\neg(\neg\varphi \vee \neg\psi)$, $\neg\varphi \vee \psi$, $(\varphi \longrightarrow \psi) \wedge (\psi \longrightarrow \varphi)$, $\neg\exists y.\neg\varphi$, and $\neg\exists X.\neg\varphi$ respectively.

In the relative distance logic, D will be allowed to be existantially quantified at the beginning of a formula. Formally, we define the **relative distance logic**, denoted by $\mathcal{L}\overleftarrow{\mathsf{d}}(\Sigma)$, to be the smallest class of formulas containing all formulas generated by the next two rules.

1. If $\varphi \in \mathsf{MSO}(T\Sigma^*)$, so is $\varphi \in \mathcal{L}\overleftarrow{\mathsf{d}}(\Sigma)$.
2. If $\varphi \in \mathcal{L}\overleftarrow{\mathsf{d}}(\Sigma)$, so is $\exists D.\varphi \in \mathcal{L}\overleftarrow{\mathsf{d}}(\Sigma)$.

Formulas of $\mathcal{L}\overleftarrow{\mathsf{d}}(\Sigma)$ are interpreted over timed words over Σ. For this, we associate with $w \in T\Sigma^*$ the relational structure consisting of the domain $\mathsf{dom}(w)$ together with the binary relation $P_a = \{i \in \mathsf{dom}(w) : a_i = a\}$ and the usual $=$ and $<$ relations on $\mathsf{dom}(w)$. We further define the binary relation $\overleftarrow{\mathsf{d}}(\cdot, \cdot) \sim c$ to be $(I, i) \in 2^{\mathsf{dom}(w)} \times \mathsf{dom}(w)$ such that one of the following conditions is satisfied

- there is some $j \in I$ such that $j < i$, $t_i - t_j \sim c$ and there is no $k \in I$ with $j < k < i$,
- there is no $j \in I$ such that $j < i$, and $t_i - 0 \sim c$.

For $\varphi \in \mathcal{L}\overleftarrow{\mathsf{d}}(\Sigma)$, let $\mathsf{Free}(\varphi)$ be the set of free variables, i.e., variables not bound by any quantifier, $\mathcal{V} \supseteq \mathsf{Free}(\varphi)$ be a finite set of first- and second-order variables, and σ be a $(\mathcal{V}, w)$-assignment mapping first-order (second-order, respectively) variables to elements (subsets, respectively) of $\mathsf{dom}(w)$. For $i \in \mathsf{dom}(w)$, we let $\sigma[y \to i]$ be the assignment that maps y to i and agrees with σ on every variable $\mathcal{V} \setminus \{y\}$. Similarly, we define $\sigma[X \to I]$ for any $I \subseteq \mathsf{dom}(w)$. For Σ, we define the extended alphabets $\Sigma_{\mathcal{V}} = \Sigma \times \{0,1\}^{\mathcal{V}}$ for every finite set $\mathcal{V}$ of variables. A timed word $w \in T\Sigma^*$ and a $(\mathcal{V}, w)$-assignment σ are encoded as timed word over the extended alphabet $\Sigma_{\mathcal{V}}$. A timed word over $\Sigma_{\mathcal{V}}$ is written as $((\bar{a}, \sigma), \bar{t})$, where $(\bar{a}, \bar{t})$ is the projection over $T\Sigma^*$ and σ is the projection over $\{0,1\}^{\mathcal{V}}$. Then, σ represents a *valid* assignment over $\mathcal{V}$ if for each first-order variable $y \in \mathcal{V}$, the y-row of σ contains exactly one 1. In this case, σ is identified with the $(\mathcal{V}, w)$-assignment such that for every first-order variable $y \in \mathcal{V}$, $\sigma(y)$ is the position of the 1 in the y-row, and for each second-order variable $X \in \mathcal{V}$, $\sigma(X)$ is the set of positions with a 1 in the X-row.

Example 5.1. Let $\Gamma = \{a, b\}$ and $w = (a, 2.0)(a, 3.5)(b, 4.2)$ be a timed word over Γ. Further let $\mathcal{V} = \{y, X\}$ and consider the valid $(\mathcal{V}, w)$-assignment σ with $\sigma(y) = 2$ and $\sigma(X) = \{1, 2\}$. We encode w and σ as the timed word $\left(\begin{smallmatrix}a\\0\\1\end{smallmatrix}, 2.0\right)\left(\begin{smallmatrix}a\\1\\1\end{smallmatrix}, 3.5\right)\left(\begin{smallmatrix}b\\0\\0\end{smallmatrix}, 4.2\right)$ over $\Gamma_{\mathcal{V}}$.

We define $N_{\mathcal{V}} = \{((\bar{a}, \sigma), \bar{t}) \in T(\Sigma_{\mathcal{V}})^* : \sigma \text{ is a valid } (\mathcal{V}, (\bar{a}, \bar{t}))\text{-assignment}\}$. The definition that $((\bar{a}, \sigma), \bar{t})$ satisfies φ, written $((\bar{a}, \sigma), \bar{t}) \models \varphi$, is as usual provided that the domain of σ contains $\mathsf{Free}(\varphi)$. We let $L_{\mathcal{V}}(\varphi) = \{((\bar{a}, \sigma), \bar{t}) \in N_{\mathcal{V}} : ((\bar{a}, \sigma), \bar{t}) \models \varphi\}$. The formula φ *defines* the timed language $L(\varphi) = L_{\mathsf{Free}(\varphi)}(\varphi)$. A formula φ is a *sentence* if $\mathsf{Free}(\varphi) = \emptyset$. A timed language $L \subseteq T\Sigma^*$ is $\mathcal{L}\overleftarrow{\mathsf{d}}(\Sigma)$*-definable* if there exists a sentence $\varphi \in \mathcal{L}\overleftarrow{\mathsf{d}}(\Sigma)$ such that $L(\varphi) = L$.

Theorem 5.2 ([107]). *A timed language $L \subseteq T\Sigma^*$ is $\mathcal{L}\overleftarrow{\mathsf{d}}(\Sigma)$-definable if and only if L is TA-recognizable over Σ. The transformations from a timed automaton over Σ to a $\mathcal{L}\overleftarrow{\mathsf{d}}(\Sigma)$-sentence and back can be done efficiently.*

Now, we turn to the weighted extension of these logics. *For this, we fix a semiring $\mathcal{K}$ and a family $\mathcal{F}$ of functions from $\mathbb{R}_{\geq 0}$ to K containing $\mathbb{1}$.* Again, we start with the **underlying auxiliary logic** and define it by the following grammar.

$$\begin{aligned}\varphi \ ::= \ & P_a(y) \mid \neg P_a(y) \mid y = z \mid \neg(y = z) \mid y < z \mid \neg(y < z) \mid y \in X \mid \neg(y \in X) \mid \\ & \overleftarrow{\mathsf{d}}(D, y) \sim c \mid \neg(\overleftarrow{\mathsf{d}}(D, y) \sim c) \mid k \mid f(y) \mid \varphi \vee \varphi \mid \varphi \wedge \varphi \mid Qy.\varphi \mid QX.\varphi\end{aligned}$$

where y, z are first-order variables, X, D are second-order variables, $Q \in \{\exists, \forall\}$, $a \in \Sigma$, $c \in \mathbb{N}$, $\sim \in \{<, \leq, =, \geq, >\}$, $k \in K$ and $f \in \mathcal{F}$. We use $\mathsf{MSO}(\mathcal{K}, T\Sigma^*, \mathcal{F})$ to denote the collection of all such formulas. Formulas of the form k and $f(y)$ are called *weighted atomic* formulas.

Notice that negation may only be applied to atomic formulas of $\mathsf{MSO}(T\Sigma^*)$. This is because for arbitrary semirings it is not clear what the negation of a weighted atomic formula should mean. In the following, we use the term *atomic formulas* to refer atomic formulas of $\mathsf{MSO}(T\Sigma^*)$ and their negations.

Finally, we define the **weighted relative distance logic**, denoted by $\mathcal{L}\overleftarrow{\mathsf{d}}(\mathcal{K}, \Sigma, \mathcal{F})$, to be the smallest class of formulas containing all formulas generated by the next two rules.

1. If $\varphi \in \mathsf{MSO}(\mathcal{K}, T\Sigma^*, \mathcal{F})$, then $\varphi \in \mathcal{L}\overleftarrow{\mathsf{d}}(\mathcal{K}, \Sigma, \mathcal{F})$.

2. If $\varphi \in \mathcal{L}\overleftarrow{\mathsf{d}}(\mathcal{K}, \Sigma, \mathcal{F})$, then $\exists D.\varphi \in \mathcal{L}\overleftarrow{\mathsf{d}}(\mathcal{K}, \Sigma, \mathcal{F})$.

Next, we define the semantics of this logic. Let $\varphi \in \mathcal{L}\overleftarrow{\mathsf{d}}(\mathcal{K}, \Sigma, \mathcal{F})$ and $\mathcal{V} \supseteq \mathsf{Free}(\varphi)$. The $\mathcal{V}$-semantics of φ is a timed series $[\![\varphi]\!]_{\mathcal{V}} : T(\Sigma_{\mathcal{V}})^* \to K$. Let $(\bar{a}, \bar{t}) \in T\Sigma^*$. If σ is a valid $(\mathcal{V}, (\bar{a}, \bar{t}))$-assignment, $\big([\![\varphi]\!]_{\mathcal{V}}, ((\bar{a}, \sigma), \bar{t})\big) \in K$ is defined inductively as follows:

$$\begin{aligned}
\big([\![\varphi]\!]_{\mathcal{V}}, ((\bar{a}, \sigma), \bar{t})\big) &= \big(1_{L_{\mathcal{V}}(\varphi)}, ((\bar{a}, \sigma), \bar{t})\big) \text{ if } \varphi \text{ is atomic} \\
\big([\![k]\!]_{\mathcal{V}}, ((\bar{a}, \sigma), \bar{t})\big) &= k \\
\big([\![f(y)]\!]_{\mathcal{V}}, ((\bar{a}, \sigma), \bar{t})\big) &= f(t_{\sigma(y)} - t_{\sigma(y)-1}) \\
\big([\![\varphi \vee \varphi']\!]_{\mathcal{V}}, ((\bar{a}, \sigma), \bar{t})\big) &= \big([\![\varphi]\!]_{\mathcal{V}}, ((\bar{a}, \sigma), \bar{t})\big) + \big([\![\varphi']\!]_{\mathcal{V}}, ((\bar{a}, \sigma), \bar{t})\big) \\
\big([\![\varphi \wedge \varphi']\!]_{\mathcal{V}}, ((\bar{a}, \sigma), \bar{t})\big) &= \big([\![\varphi]\!]_{\mathcal{V}}, ((\bar{a}, \sigma), \bar{t})\big) \cdot \big([\![\varphi']\!]_{\mathcal{V}}, ((\bar{a}, \sigma), \bar{t})\big) \\
\big([\![\exists y.\varphi]\!]_{\mathcal{V}}, ((\bar{a}, \sigma), \bar{t})\big) &= \sum_{i \in \mathsf{dom}((\bar{a}, \bar{t}))} \big([\![\varphi]\!]_{\mathcal{V} \cup \{y\}}, ((\bar{a}, \sigma[y \to i]), \bar{t})\big) \\
\big([\![\forall y.\varphi]\!]_{\mathcal{V}}, ((\bar{a}, \sigma), \bar{t})\big) &= \prod_{i \in \mathsf{dom}((\bar{a}, \bar{t}))} \big([\![\varphi]\!]_{\mathcal{V} \cup \{y\}}, ((\bar{a}, \sigma[y \to i]), \bar{t})\big) \\
\big([\![\exists X.\varphi]\!]_{\mathcal{V}}, ((\bar{a}, \sigma), \bar{t})\big) &= \sum_{I \subseteq \mathsf{dom}((\bar{a}, \bar{t}))} \big([\![\varphi]\!]_{\mathcal{V} \cup \{X\}}, ((\bar{a}, \sigma[X \to I]), \bar{t})\big) \\
\big([\![\exists D.\varphi]\!]_{\mathcal{V}}, ((\bar{a}, \sigma), \bar{t})\big) &= \sum_{I \subseteq \mathsf{dom}((\bar{a}, \bar{t}))} \big([\![\varphi]\!]_{\mathcal{V} \cup \{D\}}, ((\bar{a}, \sigma[D \to I]), \bar{t})\big)
\end{aligned}$$

$$\big(\llbracket \forall X.\varphi \rrbracket_{\mathcal{V}}, ((\bar{a},\sigma),\bar{t})\big) \quad = \prod_{I \subseteq \mathsf{dom}(\bar{a},\bar{t})} \big(\llbracket \varphi \rrbracket_{\mathcal{V}\cup\{X\}}, ((\bar{a},\sigma[X \to I]),\bar{t})\big)$$

For σ not a valid $(\mathcal{V},(\bar{a},\bar{t}))$-assignment, we define $\big(\llbracket\varphi\rrbracket_{\mathcal{V}}((\bar{a},\sigma),\bar{t})\big) = 0$. We write $\llbracket\varphi\rrbracket$ for $\llbracket\varphi\rrbracket_{\mathsf{Free}(\varphi)}$.

Remark 5.3. If $\mathcal{K}$ is the Boolean semiring, then $\mathcal{L}\overleftarrow{\mathsf{d}}(\mathcal{K},\Sigma,\mathcal{F})$ corresponds to $\mathcal{L}\overleftarrow{\mathsf{d}}(\Sigma)$. This is because every formula in $\mathcal{L}\overleftarrow{\mathsf{d}}(\Sigma)$ is language equivalent to a formula where negation is applied to atomic subformulas only. Also, every such formula in $\mathcal{L}\overleftarrow{\mathsf{d}}(\Sigma)$ can be seen as a formula of $\mathcal{L}\overleftarrow{\mathsf{d}}(\mathcal{K},\Sigma,\mathcal{F})$.

Example 5.4. Consider the formula $\varphi = \exists D.\exists y.P_b(y) \wedge \overleftarrow{\mathsf{d}}(D,y) < 2$ and let $w = (a,1.0)(a,2.0)(b,3.0)$. If $\mathcal{K}$ is the Boolean semiring or, equivalently, we interprete φ as an $\mathcal{L}\overleftarrow{\mathsf{d}}(\Sigma)$-formula, we have $(\llbracket\varphi\rrbracket, w) = 1$, as, for instance, the time difference between the third and second position is less than 2, so we may choose $\sigma(y) = 3$ and $\sigma(D)$ such that $2 \in \sigma(D)$. If on the other hand, we let $\mathcal{K}$ be the semiring of the natural numbers with ordinary addition and multiplication, we have $(\llbracket\varphi\rrbracket, w) = 4$, since there are 4 different assignments such that $P_b(y) \wedge \overleftarrow{\mathsf{d}}(D,y) < 2$ is evaluated to 1. In fact, using this semiring, we can *count* how often a certain property holds. This may give rise to interesting applications in the field of verification.

Example 5.5. We let $\mathcal{K}$ be the max-plus-semiring and $f(\delta) = \delta$ for each $\delta \in \mathbb{R}_{\geq 0}$. Then, the formula $\varphi = \exists y.f(y)$ computes for each timed word w the maximal time difference $t_i - t_{i-1}$ between two consecutive events. Formally, $(\llbracket\varphi\rrbracket, w) = \max\{t_i - t_{i-1} : i \in \mathsf{dom}(w)\}$ for each $w \in T\Sigma^*$. Note that $\llbracket\varphi\rrbracket$ is recognized by the weighted timed automaton of Ex. 3.2.

The following lemma states that for each formula φ of our logic, the semantics for different finite sets $\mathcal{V}$ of variables containing $\mathsf{Free}(\varphi)$ are consistent with each other. It can be proved by induction on the structure of $\mathcal{L}\overleftarrow{\mathsf{d}}(\mathcal{K},\Sigma,\mathcal{F})$

Lemma 5.6. *Let $\varphi \in \mathcal{L}\overleftarrow{\mathsf{d}}(\mathcal{K},\Sigma,\mathcal{F})$ and $\mathcal{V}$ a finite set of variables containing* $\mathsf{Free}(\varphi)$. *Then*

$$\big(\llbracket\varphi\rrbracket_{\mathcal{V}}, ((\bar{a},\sigma),\bar{t})\big) = \big(\llbracket\varphi\rrbracket, ((\bar{a},\sigma_{|\mathsf{Free}(\varphi)}),\bar{t})\big)$$

for each $((\bar{a},\sigma),\bar{t}) \in T(\Sigma_{\mathcal{V}})^*$ such that σ is a valid $(\mathcal{V},(\bar{a},\bar{t}))$-assignment.

Let $\mathfrak{L} \subseteq \mathcal{L}\overleftarrow{\mathsf{d}}(\mathcal{K},\Sigma,\mathcal{F})$. A timed series $\mathcal{T} : T\Sigma^* \to K$ is called $\mathfrak{L}$-*definable* if there is a sentence $\varphi \in \mathfrak{L}$ such that $\llbracket\varphi\rrbracket = \mathcal{T}$. The goal of this section is to find a suitable fragment $\mathfrak{L} \subseteq \mathcal{L}\overleftarrow{\mathsf{d}}(\mathcal{K},\Sigma,\mathcal{F})$ such that $\mathfrak{L}$-definable timed series precisely correspond to $\mathcal{F}$-recognizable timed series. In other words, we want to generalize Theorem 5.2 to the weighted setting. It is not surprising that $\mathcal{L}\overleftarrow{\mathsf{d}}(\mathcal{K},\Sigma,\mathcal{F})$ itself does not constitute a suitable candidate for $\mathfrak{L}$, since also in the untimed setting, the full weighted MSO logics is expressively stronger than weighted finite automata [49]. In the next section, we explain the problems that occur when we do not restrict the logic. For simplicity, we do this exemplarily for the case of *idempotent and commutative* semirings. To be as general as possible, we will moreover consider families of functions that are *not closed under pointwise product*. Notice that this setting includes weighted timed automata over the max-plus-semiring and the family of functions of the form δ^k for some $k \in \mathbb{N}$ and each $\delta \in \mathbb{R}_{\geq 0}$. Stepwisely, we develop solutions resulting in a fragment of $\overleftarrow{\mathcal{L}\mathsf{d}}(\mathcal{K}, \Sigma, \mathcal{F})$ for which we are able to present a Büchi theorem for weighted timed automata over this particular setting. Later we will show how to generalize this approach to arbitrary semirings.

5.3 From Definability to Recognizability

In this section, we fix an idempotent and commutative semiring $\mathcal{K}$. Moreover, we assume that $\mathcal{F}$ is not necessarily closed under pointwise product.

We want to develop a fragment $\mathfrak{L} \subseteq \overleftarrow{\mathcal{L}\mathsf{d}}(\mathcal{K}, \Sigma, \mathcal{F})$ such that for every sentence $\varphi \in \mathfrak{L}$, $[\![\varphi]\!]$ is an $\mathcal{F}$-recognizable timed series. As in the classical setting, the proof for this is done by induction over the structure of the logic: for the induction base, we show that for every atomic formula φ in $\overleftarrow{\mathcal{L}\mathsf{d}}(\mathcal{K}, \Sigma, \mathcal{F})$, there is a weighted timed automaton $\mathcal{A}$ over $\mathcal{K}$, $\Sigma_{\mathsf{Free}(\varphi)}$[1] and $\mathcal{F}$ such that $\|\mathcal{A}\| = [\![\varphi]\!]$. For the induction step, we need to show that $\mathcal{F}$-recognizable timed series are closed under the operators of $\mathfrak{L}$. In the case of disjunction and existential quantification, the proofs are very similar to the classical case [103, 49]. In the case of conjunction and universal quantification, however, problems arise. Problems with unrestricted use of conjunctions are due to the fact that $\mathcal{F}$-recognizable timed series are not closed under Hadamard product in general (see Example 3.4). Problems with unrestricted use of universal quantification are due to the fact that the semantics of formulas may grow too fast with the size of a timed word to be recognizable by a weighted timed automaton. This is demonstrated in the next example.

Example 5.7. Let $\mathcal{K}$ be the max-plus-semiring and $\mathcal{F}$ be the family of functions of the form δ^k for some $k \in \mathbb{N}$ and all $\delta \in \mathbb{R}_{\geq 0}$. We let f be the function defined by $f(\delta) = \delta^1$ for each $\delta \in \mathbb{R}_{\geq 0}$. We consider the formula $\varphi = \forall z.\exists y.f(y)$. Then we have $([\![\varphi]\!], w) = |w| \cdot \max\{t_i - t_{i-1} : i \in \mathsf{dom}(w)\}$ for each $w \in T\Sigma^*$. However, $[\![\varphi]\!]$ is not $\mathcal{F}$-recognizable, as is proved in the following: assume $\mathcal{A} = (\mathcal{L}, \mathcal{C}, E, \mathsf{in}, \mathsf{out}, \mathsf{ewt}, \mathsf{lwt})$ is a weighted timed automaton over $\mathcal{K}$, Σ and $\mathcal{F}$ such that $\|\mathcal{A}\| = [\![\varphi]\!]$. Notice that the weight functions in, out and ewt assign constants to the locations and edges, respectively. Thus, for each location l, there is some $\delta \in \mathbb{R}_{\geq 0}$ such that $\mathsf{lwt}(l)(\delta)$ is strictly greater than each

[1] Notice that atomic and weighted atomic formulas may contain free variables. Thus the weighted timed automata recognizing the semantics of an atomic formula φ are defined over the extended alphabet $\Sigma_{\mathsf{Free}(\varphi)}$.

of these constants. For this reason, we may assume $\mathsf{in}(l) = \mathsf{out}(l) = 0$ for each $l \in \mathcal{L}$ and $\mathsf{ewt}(e) = 0$ for each $e \in E$. For each $l \in \mathcal{L}$, we use c_l to denote the constant to which power the time delay in l is taken of, i.e., if $\mathsf{lwt}(l)(\delta) = \delta^n$ for each $\delta \in \mathbb{R}_{\geq 0}$ and some $n \in K$, then, $c_l = n$. Let $M = \max\{c_l : l \in \mathcal{L}\}$. Then, for every timed word $w \in T\Sigma^*$ and for each run of $\mathcal{A}$ on w we have $\mathsf{rwt}(r) \leq \sum_{1 \leq i \leq |w|}(t_i - t_{i-1})^M$. Furthermore, we have $(\|\mathcal{A}\|, w) = \max\{\mathsf{rwt}(r) : r \text{ is a run of } \mathcal{A} \text{ on } w\}$ and thus $(\|\mathcal{A}\|, w) \leq \sum_{1 \leq i \leq |w|}(t_i - t_{i-1})^M$. Now choose $w \in T\Sigma^*$ such that $|w| > 2^M$ and there exists some $i \in \mathsf{dom}(w)$ such that $t_i - t_{i-1} = 2$ and for all $j \in \mathsf{dom}(w)$ with $j \neq i$ we have $t_j - t_{j-1} = 0$. Then we obtain $(\|\mathcal{A}\|, w) < [\![\varphi]\!]$, a contradiction. The timed series $[\![\varphi]\!]$ grows too fast to be $\mathcal{F}$-recognizable.

Similar examples can be given for $\forall X$. It turns out that we have to restrict the application of these operators syntactically to show that they preserve $\mathcal{F}$-recognizability of timed series. We restrict the application of conjunction and universal quantification and consider a *syntactically restriced* fragment of $\mathsf{MSO}(\mathcal{K}, T\Sigma^*, \mathcal{F})$. We present the definition of this fragment in the following. We start with the definition of *unweighted* and *almost unambiguous* formulas.

We say that a formula $\varphi \in \mathsf{MSO}(\mathcal{K}, T\Sigma^*, \mathcal{F})$ is *unweighted*, if it does not contain any weighted atomic formulas. It can be easily seen, that unweighted formulas are in $\mathsf{MSO}(T\Sigma^*)$.

Let y be a first-order variable. We say that a formula $\psi \in \mathsf{MSO}(\mathcal{K}, T\Sigma^*, \mathcal{F})$ is *almost unambiguous over* y, if it is in the disjunctive and conjunctive closure of unweighted formulas, constants $k \in K$ and formulas $f(y)$ for some $f \in \mathcal{F}$, such that $f(y)$ may appear at most once in every subformula of ψ of the form $\psi_1 \wedge \psi_2$.

Given a formula $\varphi \in \mathsf{MSO}(\mathcal{K}, T\Sigma^*, \mathcal{F})$, we define the set $\mathcal{V}_f(\varphi)$ to be the set of all first-order variables y such that $f(y)$ appears in φ.

We define the **syntactically restricted auxiliary logic** $\mathsf{sRMSO}(\mathcal{K}, T\Sigma^*, \mathcal{F})$ to be the smallest class of formulas generated by the following rules.

1. If $\varphi \in \mathsf{MSO}(\mathcal{K}, T\Sigma^*, \mathcal{F})$ is an atomic or a weighted atomic formula, then $\varphi \in \mathsf{sRMSO}(\mathcal{K}, T\Sigma^*, \mathcal{F})$.
2. If $\varphi, \psi \in \mathsf{sRMSO}(\mathcal{K}, T\Sigma^*, \mathcal{F})$, then $\varphi \vee \psi, \exists y.\varphi, \exists X.\varphi \in \mathsf{sRMSO}(\mathcal{K}, T\Sigma^*, \mathcal{F})$.
3. If $\varphi \in \mathsf{sRMSO}(\mathcal{K}, T\Sigma^*, \mathcal{F})$ is unweighted, then $\forall X.\varphi \in \mathsf{sRMSO}(\mathcal{K}, T\Sigma^*, \mathcal{F})$.
4. If $\varphi \in \mathsf{sRMSO}(\mathcal{K}, T\Sigma^*, \mathcal{F})$ is almost unambiguous over y, then $\forall y.\varphi \in \mathsf{sRMSO}(\mathcal{K}, T\Sigma^*, \mathcal{F})$.
5. If $\varphi, \psi \in \mathsf{sRMSO}(\mathcal{K}, T\Sigma^*, \mathcal{F})$ and at least one of the following conditions hold
   - $\mathcal{V}_f(\varphi) = \emptyset$,
   - $\mathcal{V}_f(\psi) = \emptyset$,

- $\mathcal{V}_f(\varphi) \cap \mathcal{V}_f(\psi) = \emptyset$, $\mathcal{V}_f(\varphi) \subseteq \mathsf{Free}(\varphi)$, and $\mathcal{V}_f(\psi) \subseteq \mathsf{Free}(\psi)$,

then $\varphi \wedge \psi \wedge \bigwedge_{\substack{y \in \mathcal{V}_f(\varphi), z \in \mathcal{V}_f(\psi) \\ y \neq z}} \neg(y = z) \in \mathsf{sRMSO}(\mathcal{K}, T\Sigma^*, \mathcal{F})$.

Remark 5.8. If $\mathcal{F}$ is closed under pointwise product, we can replace condition 5 by the following rule: if $\varphi, \psi \in \mathsf{sRMSO}(\mathcal{K}, T\Sigma^*, \mathcal{F})$, then $\varphi \wedge \psi \in \mathsf{sRMSO}(\mathcal{K}, T\Sigma^*, \mathcal{F})$.

Now, we want to show that for each formula $\varphi \in \mathsf{sRMSO}(\mathcal{K}, T\Sigma^*, \mathcal{F})$, there is a weighted timed automaton $\mathcal{A}$ such that $\|\mathcal{A}\| = [\![\varphi]\!]$. As mentioned before, this is done by induction. However, due to our restriction on conjunction in $\mathsf{sRMSO}(\mathcal{K}, T\Sigma^*, \mathcal{F})$, we will prove an even stronger result, stated in the next theorem.

Given a formula $\varphi \in \mathsf{MSO}(\mathcal{K}, T\Sigma^*, \mathcal{F})$, we let $\mathsf{Func}(\varphi)$ be the set of functions $f \in \mathcal{F}$ such that φ contains a subformula $f(y)$ for some first-order variable y. Given a weighted timed automaton $\mathcal{A}$, we let $\mathsf{Func}(\mathcal{A})$ be the set of functions f such that $\mathsf{lwt}(l) = f$ for some location l in $\mathcal{A}$.

Theorem 5.9. *Let $\varphi \in \mathsf{sRMSO}(\mathcal{K}, T\Sigma^*, \mathcal{F})$ be a syntactically restricted formula. Then for each finite set $\mathcal{V} \supseteq \mathsf{Free}(\varphi)$ there is some weighted timed automaton $\mathcal{A}_\varphi$ over $\mathcal{K}$, $\Sigma_\mathcal{V}$ and $\mathcal{F}$ such that*

1. $\|\mathcal{A}_\varphi\| = [\![\varphi]\!]_\mathcal{V}$,
2. $\mathsf{Func}(\mathcal{A}_\varphi) \subseteq \mathsf{Func}(\varphi) \cup \{\mathbb{1}\}$,
3. *for each formula $f(y)$ occurring in φ with $y \in \mathsf{Free}(\varphi)$, whenever $\mathsf{lwt}(l) = f$ for some location l in $\mathcal{A}_\varphi$, then for each edge $(l, (a, \sigma), \phi, \lambda, l')$ in $\mathcal{A}_\varphi$ we have $\sigma(y) = 1$.*

The remainder of this section is devoted to the proof of this theorem. Before, we present some notations and lemmas.

Lemma 5.10. *The semantics $[\![\varphi]\!]$ of each unweighted formula $\varphi \in \mathsf{MSO}(\mathcal{K}, T\Sigma^*, \mathcal{F})$ takes only values in $\{0, 1\}$. In particular, $[\![\varphi]\!] = 1_{L(\varphi)}$.*

PROOF. Follows from semiring axioms and idempotence of $\mathcal{K}$. ■

We say that two formulas ψ and ζ are *equivalent*, written $\psi \equiv \zeta$, if $[\![\psi]\!]_{\mathsf{Free}(\psi) \cup \mathsf{Free}(\zeta)} = [\![\zeta]\!]_{\mathsf{Free}(\psi) \cup \mathsf{Free}(\zeta)}$. The next lemma can be proved using the semiring axioms and commutativity of $\mathcal{K}$ and Lemma 5.10.

Lemma 5.11. *Let y be a first-order variable, $k_1, k_2 \in K$, and $\psi_1, \psi_2, \psi_3 \in \mathsf{MSO}(\mathcal{K}, T\Sigma^*, \mathcal{F})$ be unweighted. Then, the following equivalences hold:*

1. $\psi_1 \wedge (\psi_2 \vee \psi_3) \equiv (\psi_1 \wedge \psi_2) \vee (\psi_1 \wedge \psi_3)$
2. $\psi_1 \wedge \psi_2 \equiv \psi_2 \wedge \psi_1$

3. $\psi_1 \equiv 1 \wedge \psi_1$

4. $\psi_1 \equiv \mathbb{1}(y) \wedge \psi_1$

5. $\psi_1 \equiv \psi_1 \wedge \mathsf{true}$

6. $k_1 \wedge k_2 \equiv k_1 \cdot k_2$

Lemma 5.12. *Let y be a first-order variable and $\psi \in \mathsf{MSO}(\mathcal{K}, T\Sigma^*, \mathcal{F})$ be almost unambiguous over y. Then there is a formula $\zeta \in \mathsf{MSO}(\mathcal{K}, T\Sigma^*, \mathcal{F})$ such that ζ is of the form $\bigvee_{1\leq i\leq n} f_i(y) \wedge k_i \wedge \psi_i$ for some $n \in \mathbb{N}$, $f_i \in \mathcal{F}$, $k_i \in K$ and unweighted $\psi_i \in \mathsf{MSO}(\mathcal{K}, T\Sigma^*, \mathcal{F})$ for each $i \in \{1, ..., n\}$, and $\zeta \equiv \psi$.*

PROOF. We transform every almost unambiguous formula into the appropriate form using Lemma 5.11. First, transform every formula of the form $\psi_1 \wedge (\psi_2 \vee \psi_3)$ into the form $(\psi_1 \wedge \psi_2) \vee (\psi_1 \wedge \psi_3)$ and every formula of the form $(\psi_1 \vee \psi_2) \wedge \psi_3$ into the form $(\psi_1 \wedge \psi_3) \vee (\psi_2 \wedge \psi_3)$. Then, each disjunct can be put in the right form using Lemma 5.11. ∎

Let $\mathcal{V}, \mathcal{V}'$ be two finite sets of first- and second-order variables such that $\mathcal{V} \subseteq \mathcal{V}'$. Let $\pi : \Sigma_{\mathcal{V}'} \to \Sigma_{\mathcal{V}}$ be a projection defined by $(a, \sigma') \mapsto (a, \sigma'_{|\mathcal{V}})$.

Lemma 5.13. *For each weighted timed automaton $\mathcal{A}$ over $\Sigma_{\mathcal{V}}$, there is a weighted timed automaton $\mathcal{A}_{\mathcal{V}'}$ over $\Sigma_{\mathcal{V}'}$ such that*

1. $\|\mathcal{A}_{\mathcal{V}'}\| = \bar{\pi}^{-1}(\|\mathcal{A}\|) \odot 1_{N_{\mathcal{V}'}}$,

2. $\mathsf{Func}(\mathcal{A}_{\mathcal{V}'}) = \mathsf{Func}(\mathcal{A})$,

3. *for each $f \in \mathcal{F}$ and each first-order variable $y \in \mathcal{V}$, if each edge of the form $(l, (a, \sigma), \phi, \lambda, l')$ in $\mathcal{A}$ with $\mathsf{lwt}(l) = f$ satisfies $\sigma(y) = 1$, then each edge of the form $(l_1, (b, \sigma'), \phi', \lambda', l_2)$ in $\mathcal{A}_{\mathcal{V}'}$ with $\mathsf{lwt}'(l_1) = f$ satisfies $\sigma'(y) = 1$.*

PROOF. Let $\mathcal{A} = (\mathcal{L}, \mathcal{C}, E, \mathsf{in}, \mathsf{out}, \mathsf{ewt}, \mathsf{lwt})$ be a weighted timed automaton over $\Sigma_{\mathcal{V}}$. We define $\mathcal{A}' = (\mathcal{L}, \mathcal{C}, E', \mathsf{in}, \mathsf{out}, \mathsf{ewt}', \mathsf{lwt})$ over $\Sigma_{\mathcal{V}'}$ as follows. For each edge $e \in E$ of the form $(l, (a, \sigma), \phi, \lambda, l')$ with $(a, \sigma) \in \Sigma_{\mathcal{V}}$, for each $y \in \mathcal{V}' \backslash \mathcal{V}$ and for each $i \in \{0, 1\}$, there is an edge $e' \in E'$ of the form $(l, (a, \sigma'), \phi, \lambda, l')$ with $(a, \sigma') \in \Sigma_{\mathcal{V}'}$, where

$$\sigma'(z) = \begin{cases} \sigma(z) & \text{if } z \in \mathcal{V}, \\ i & \text{otherwise.} \end{cases}$$

Moreover, $\mathsf{ewt}'(e') = \mathsf{ewt}(e)$. There are no other edges in E'. Let $\mathcal{A}_{N_{\mathcal{V}'}}$ be a weighted timed automaton over $\Sigma_{\mathcal{V}'}$ with $\|\mathcal{A}_{N_{\mathcal{V}'}}\| = 1_{N_{\mathcal{V}'}}$ and $\mathsf{Func}(\mathcal{A}_{N_{\mathcal{V}'}}) = \{\mathbb{1}\}$. In this way, $\mathcal{A}_{N_{\mathcal{V}'}}$ is non-interfering with any weighted timed automaton. Now, let $\mathcal{A}_{\mathcal{V}'}$ be the product automaton of $\mathcal{A}'$ and $\mathcal{A}_{N_{\mathcal{V}'}}$ as defined in the proof of Lemma 3.5. It is straightforward to show that $\mathcal{A}_{\mathcal{V}'}$ satisfies conditions 1. to 3. ∎

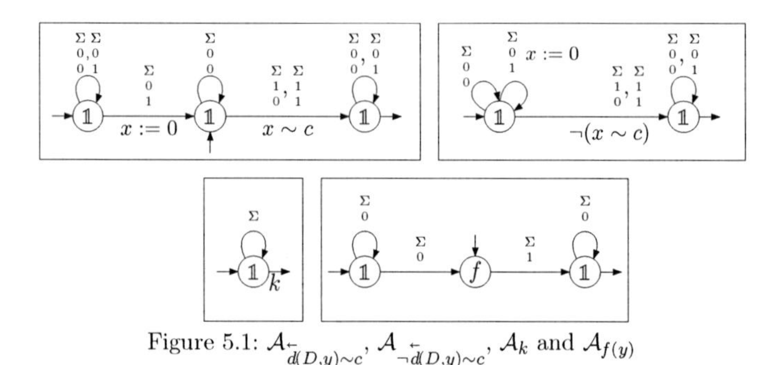

Figure 5.1: $\mathcal{A}_{\overleftarrow{d}(D,y)\sim c}$, $\mathcal{A}_{\neg\overleftarrow{d}(D,y)\sim c}$, $\mathcal{A}_k$ and $\mathcal{A}_{f(y)}$

Now, we prove Theorem 5.9. The proof is done by induction on the construction of $\mathsf{sRMSO}(\mathcal{K}, T\Sigma^*, \mathcal{F})$-formulas. We remark that we only show the claim for $\mathcal{V} = \mathsf{Free}(\varphi)$. For each other finite set $\mathcal{V}'$ of variables with $\mathcal{V}' \supseteq \mathsf{Free}(\varphi)$, the claim follows from Lemma 5.13 together with the fact that $[\![\varphi]\!]_{\mathcal{V}'} = \bar{\pi}^{-1}([\![\varphi]\!]) \odot 1_{N_{\mathcal{V}'}}$. For the induction base, we consider the atomic and weighted atomic formulas in $\mathsf{sRMSO}(\mathcal{K}, T\Sigma^*, \mathcal{F})$.

Atomic and Weighted Atomic Formulas Let $\varphi \in \mathsf{sRMSO}(\mathcal{K}, T\Sigma^*, \mathcal{F})$ be atomic. If φ equals $P_a(y)$, $y < z$, $y = z$, $y \in X$, or one of its negations, we can construct a timed automaton $\mathcal{A}'_\varphi$ using the same approach as for formulas in $\mathsf{MSO}(\Sigma)$ (see e.g. Thomas [102]). We define $\mathcal{A}_\varphi$ to be the weighted timed automaton obtained from $\mathcal{A}'_\varphi$ by adding weight functions in, out, ewt and lwt defined by $\mathsf{in}(l) = \mathsf{out}(l) = 1$ and $\mathsf{lwt}(l) = \mathbb{1}$ for each location l and $\mathsf{ewt}(e) = 1$ for each edge e. For φ not as above, the corresponding weighted timed automata $\mathcal{A}_\varphi$ are shown in Fig. 5.1. The idea behind the construction of $\mathcal{A}_\varphi$ for $\varphi = \overleftarrow{\mathsf{d}}(D, y) \sim c$ is as follows: $\mathcal{A}_\varphi$ non-deterministically guesses when the last edge labeled with a letter in $\Sigma_{\{y,D\}}$ with a 1 in the D-row is taken and resets the clock variable at this edge. Then it verifies that whenever an edge is labeled by a letter with a 1 in the y-row, the time distance to the last event labeled with a letter with a 1 in the D-row satisfies $\sim c$. This can be done by adding a corresponding clock constraint to this edge. The idea for $\varphi = \neg\overleftarrow{\mathsf{d}}(D, y) \sim c$ is similar. If $\varphi = f(y)$, $\mathcal{A}_\varphi$ verifies that whenever an edge is labeled with a letter such that there is a 1 in the y-row, then the source location of this edge must be assigned the weight function f. All the other locations must be assigned the weight function $\mathbb{1}$. Finally, it can be shown in a straightforward manner that conditions 2. and 3. of Theorem 5.9 are satisfied.

Remark 5.14. In a previous paper [53], we defined weighted timed automata with initial and final locations, i.e., we restricted the image of the weight functions in and out to $\{0, 1\}$. However, it turns out that for this model it is not possible to find a weighted timed automaton whose semantics corresponds to the formula k. The problems are due to the empty timed word, for which one cannot construct a weighted timed automaton with the appropriate behaviour in the general case, i.e., for each semiring and each family of functions.

For the induction step, we have to consider all operators of our logic. We start with disjunction and existential quantification.

Disjunction Let $\psi, \zeta \in \mathsf{sRMSO}(\mathcal{K}, T\Sigma^*, \mathcal{F})$ and assume $\varphi = \psi \vee \zeta$. Note that $\mathsf{Free}(\varphi) = \mathsf{Free}(\psi) \cup \mathsf{Free}(\zeta)$ and thus $\mathsf{Free}(\psi) \subseteq \mathsf{Free}(\varphi)$ and $\mathsf{Free}(\zeta) \subseteq \mathsf{Free}(\varphi)$. By induction hypothesis, there are weighted timed automata $\mathcal{A}_\psi$ over $\Sigma_{\mathsf{Free}(\varphi)}$ and $\mathcal{A}_\zeta$ over $\Sigma_{\mathsf{Free}(\varphi)}$, respectively, satisfying condition 1. to 3. of Theorem 5.9. Let $\mathcal{A}_\varphi$ be the weighted timed automaton obtained from $\mathcal{A}_\psi$ and $\mathcal{A}_\zeta$ as defined in the proof of Lemma 3.3. Hence, we have $\|\mathcal{A}_\varphi\| = \|\mathcal{A}_\psi\| + \|\mathcal{A}_\zeta\|$ and thus $\|\mathcal{A}_\varphi\| = [\![\varphi]\!]$. Clearly, also conditions 2. and 3. hold.

Existential quantification Let $\psi \in \mathsf{sRMSO}(\mathcal{K}, T\Sigma^*, \mathcal{F})$ and assume $\varphi = \exists y.\psi$. We further let $\mathcal{V} = \mathsf{Free}(\varphi)$ and $\mathcal{V}' = \mathcal{V} \cup \{y\} = \mathsf{Free}(\psi)$. By induction hypothesis, there is a weighted timed automaton $\mathcal{A}_\psi$ over $\Sigma_{\mathcal{V}'}$ satisfying conditions 1. to 3. of Theorem 5.9. Let $p : \Sigma_{\mathcal{V}'} \to \Sigma_\mathcal{V}$ be the projection that simply erases the y-row. Let $\mathcal{A}_\varphi$ be the weighted timed automaton over $\Sigma_\mathcal{V}$ obtained from $\mathcal{A}_\psi$ as defined in the proof of Lemma 3.7. Hence, we have $\|\mathcal{A}_\varphi\| = \bar{p}(\|\mathcal{A}_\psi\|)$. However, for each $((\bar{a}, \sigma), \bar{t}) \in T(\Sigma_\mathcal{V})^*$, we also have

$$
\begin{aligned}
& \big(\bar{p}(\|\mathcal{A}_\psi\|), ((\bar{a}, \sigma), \bar{t})\big) \\
= \; & \big(\bar{p}([\![\psi]\!]_{\mathcal{V}\cup\{y\}}), ((\bar{a}, \sigma), \bar{t})\big) \\
= \; & \sum_{\substack{((\bar{a},\sigma'),\bar{t}) \in T(\Sigma_{\mathcal{V}\cup\{y\}})^* \\ p((\bar{a},\sigma'),\bar{t}) = ((\bar{a},\sigma),\bar{t})}} \big([\![\psi]\!]_{\mathcal{V}\cup\{y\}}, ((\bar{a}, \sigma'), \bar{t})\big) \\
\stackrel{\star}{=} \; & \sum_{i \in \mathsf{dom}(\bar{a},\bar{t})} \big([\![\psi]\!]_{\mathcal{V}\cup\{y\}}, ((\bar{a}, \sigma[y \to i], \bar{t})\big) \\
= \; & \big([\![\exists y.\psi]\!]_\mathcal{V}, ((\bar{a}, \sigma), \bar{t})\big)
\end{aligned}
$$

where * uses the equivalences

$$p\big((\bar{a}, \sigma'), \bar{t}\big) = ((\bar{a}, \sigma), \bar{t}) \quad \Leftrightarrow \quad \sigma' = \sigma[y \to i] \text{ for some } i \in \mathsf{dom}(\bar{a}, \bar{t})$$

and

$$\sigma \text{ is a valid } (\mathcal{V}, (\bar{a}, \bar{t}))\text{-assignment}$$

$$\Leftrightarrow$$

$$\sigma[y \to i] \text{ is a valid } (\mathcal{V} \cup \{y\}, (\bar{a}, \bar{t}))\text{-assignment for every } i \in \mathsf{dom}(\bar{a}, \bar{t}).$$

Hence, $\|\mathcal{A}_\varphi\| = [\![\varphi]\!]$. It is obvious that condition 2. is satisfied. For showing condition 3., let $f(z)$ be a subformula occuring in φ with $z \in \mathsf{Free}(\varphi)$. Thus, $z \neq y$. Since $\mathcal{A}_\varphi$ is obtained from $\mathcal{A}_\psi$ by only removing the y-row from the labels of all edges, condition 3. is satisfied.

The proof for the case $\varphi = \exists X.\psi$ can be done analogously.

Conjunction Let $\psi, \zeta \in \mathsf{sRMSO}(\mathcal{K}, T\Sigma^*, \mathcal{F})$ and assume that one of the following conditions hold:

- $\mathcal{V}_f(\psi) = \emptyset$,
- $\mathcal{V}_f(\zeta) = \emptyset$,
- $\mathcal{V}_f(\psi) \cap \mathcal{V}_f(\zeta) = \emptyset$, $\mathcal{V}_f(\psi) \subseteq \mathsf{Free}(\psi)$, and $\mathcal{V}_f(\zeta) \subseteq \mathsf{Free}(\zeta)$,

Further assume $\varphi = \psi \wedge \zeta \wedge \bigwedge_{\substack{y \in \mathcal{V}_f(\psi), z \in \mathcal{V}_f(\zeta) \\ y \neq z}} \neg(y = z)$.

Note that $\mathsf{Free}(\varphi) = \mathsf{Free}(\psi) \cup \mathsf{Free}(\zeta)$ and thus $\mathsf{Free}(\psi) \subseteq \mathsf{Free}(\varphi)$ and $\mathsf{Free}(\zeta) \subseteq \mathsf{Free}(\varphi)$. By induction hypothesis, there are weighted timed automata $\mathcal{A}_\psi$ over $\Sigma_{\mathsf{Free}(\varphi)}$ and $\mathcal{A}_\zeta$ over $\Sigma_{\mathsf{Free}(\varphi)}$ satisfying conditions 1. to 3. of Theorem 5.9.

We now distinguish between three cases.

(Case 1) We assume that $\mathcal{V}_f(\psi) = \emptyset$. Hence, $\mathsf{Func}(\psi) = \emptyset$. By induction hypothesis, we thus have $\mathsf{Func}(\mathcal{A}_\psi) = \{\mathbb{1}\}$. This implies that $\mathcal{A}_\psi$ is non-interfering with $\mathcal{A}_\zeta$. We let $\mathcal{A}_\varphi$ be the product automaton of $\mathcal{A}_\psi$ and $\mathcal{A}_\zeta$ as defined in the proof of Lemma 3.5. Hence, we have $\|\mathcal{A}_\varphi\| = \|\mathcal{A}_\psi\| \odot \|\mathcal{A}_\zeta\|$ and thus $\|\mathcal{A}_\varphi\| = [\![\varphi]\!]$. It is straightforward to show that $\mathcal{A}_\varphi$ also satisfies conditions 2. and 3.

(Case 2) We assume that $\mathcal{V}_f(\zeta) = \emptyset$. This case be done analogously to case 1.

(Case 3) Assume that both $\mathcal{V}_f(\psi) \neq \emptyset$ and $\mathcal{V}_f(\zeta) \neq \emptyset$, and thus we may assume

(a) $\mathcal{V}_f(\psi) \cap \mathcal{V}_f(\zeta) = \emptyset$,

(b) $\mathcal{V}_f(\psi) \subseteq \mathsf{Free}(\psi)$, and

(c) $\mathcal{V}_f(\zeta) \subseteq \mathsf{Free}(\zeta)$.

Let $\chi = \bigwedge_{\substack{y \in \mathcal{V}_f(\psi), z \in \mathcal{V}_f(\zeta) \\ y \neq z}} \neg(y = z)$ and put $\varrho = \zeta \wedge \chi$. Since $\mathcal{V}_f(\chi) = \emptyset$, the conditions of case 2 are satisfied and hence there is a weighted timed automaton $\mathcal{A}_\varrho$ over $\Sigma_{\mathsf{Free}(\varphi)}$ satisfying conditions 1. to 3.

Next, we show that $\mathcal{A}_\psi$ and $\mathcal{A}_\varrho$ are non-interfering. For this, let l be a location in $\mathcal{A}_\psi$ such that $\mathsf{lwt}_\psi(l) = f$ for some $f \in \mathcal{F}$. By condition 2. of Theorem 5.9, there is some subformula $f(y)$ occurring in ψ for some first-order variable y. By (b), we know that $y \in \mathsf{Free}(\psi)$, and thus by condition 3. of Theorem 5.9, for each edge $(l,(a,\sigma),\phi,\lambda,l_1)$ in $\mathcal{A}_\psi$ we have $\sigma(y)=1$.

Now, let l' be a location in $\mathcal{A}_\varrho$ such that $\mathsf{lwt}_\varrho(l') = f'$ for some $f' \in \mathcal{F}$. By condition 2. of Theorem 5.9, there is some subformula $f'(z)$ occurring in ϱ for some first-order variable z. Clearly, by definition of ϱ, this subformula $f'(z)$ can only occur in ζ. Let $(l',(b,\sigma),\phi',\lambda',l_2)$ be an edge of $\mathcal{A}_\varrho$. By (b), we have $z \in \mathsf{Free}(\zeta) \subseteq \mathsf{Free}(\varrho)$, and thus by condition 3. of Theorem 5.9, we have $\sigma(z)=1$. We further know by (a) that $y \neq z$, which implies $\sigma(y)=0$. From this it follows that for l and l', there is no edge labeled with a common letter in $\Sigma_{\mathsf{Free}(\varphi)}$. Hence, from (l,l') there is no run into $\mathcal{L}_f$ (see Sect. 3), and thus $\mathcal{A}_\psi$ and $\mathcal{A}_\varrho$ are non-interfering.

Finally, let $\mathcal{A}_\varphi$ be the product automaton of $\mathcal{A}_\psi$ and $\mathcal{A}_\varrho$ as defined in the proof of Lemma 3.5. Clearly, we have

$$\|\mathcal{A}_\varphi\| = [\![\psi \wedge \zeta \wedge \bigwedge_{\substack{y\in\mathcal{V}_f(\psi),z\in\mathcal{V}_f(\zeta)\\ y\neq z}} \neg(y=z)]\!].$$

It is straightforward to show that conditions 2. and 3. also hold.

Second-Order Universal Quantification Now, let $\psi \in \mathsf{MSO}(\mathcal{K}, T\Sigma^*, \mathcal{F})$ be unweighted and assume $\varphi = \forall X.\psi$. Clearly, φ is also unweighted. By Lemma 5.10, $[\![\varphi]\!] = 1_{L(\varphi)}$. By Theorem 5.2, there is a timed automaton $\mathcal{A}$ such that $L(\mathcal{A}) = L(\varphi)$. Let $\mathcal{A}_\varphi$ be the weighted timed automaton obtained from $\mathcal{A}$ as defined in the proof of Lemma 3.9.2. Then $\mathcal{A}_\varphi$ satisfies conditions 1. to 3. of Theorem 5.9.

Before we come to first-order universal quantification, we introduce a normalization technique and some notations.

Lemma 5.15. *For every TA-recognizable timed language $L \subseteq T\Sigma^*$, there is a timed automaton $\mathcal{A}'$ such that $L(\mathcal{A}') = L$, and for each location l in $\mathcal{A}'$ there is a unique $a \in \Sigma$ such that every edge (l,a',ϕ,λ,l') in $\mathcal{A}'$ satisfies $a' = a$.*

PROOF. Let $L \subseteq T\Sigma^*$ be TA-recognizable over Σ. Then there is a timed automaton $\mathcal{A} = (\mathcal{L}, \mathcal{L}_0, \mathcal{L}_f, \mathcal{C}, E)$ such that $L(\mathcal{A}) = L$. Define $\mathcal{A}' = (\mathcal{L}', \mathcal{L}'_0, \mathcal{L}'_f, \mathcal{C}, E')$, where

- $\mathcal{L}' = \mathcal{L} \times \Sigma$,
- $\mathcal{L}'_0 = \mathcal{L}_0 \times \Sigma$,
- $\mathcal{L}'_f = \mathcal{L}_f \times \Sigma$,
- $E' = \{((l,a),a,\phi,\lambda,(l',a')) : (l,a,\phi,\lambda,l') \in E, a' \in \Sigma\}$.

Then we have $L(\mathcal{A}') = L(\mathcal{A})$, which can be proved in a straightforward way. ■

Let $n \in \mathbb{N}\setminus\{0\}$. We define $\Sigma^{(n)} = \Sigma \times \{1, ..., n\}$. Similarly to timed words over extended alphabets of the form $\Sigma_{\mathcal{V}}$ for some finite set $\mathcal{V}$ of variables, we write $((\bar{a}, \mu), \bar{t})$ to denote a timed word over $\Sigma^{(n)}$, where $(\bar{a}, \bar{t}) \in T\Sigma^*$ and $\mu \in \{1, ..., n\}^{\mathsf{dom}(\bar{a},\bar{t})}$. We define for every $\xi \in \mathsf{MSO}(T\Sigma^*)$ the formula $\widetilde{\xi} \in \mathsf{MSO}(T(\Sigma^{(n)})^*)$ by replacing in ξ every occurence of $P_a(y)$ by $\bigvee_{1 \leq j \leq n} P_{(a,j)}(y)$.

Lemma 5.16. *Let $\xi \in \mathsf{MSO}(T\Sigma^*)$ and $\mathcal{V} \supseteq \mathsf{Free}(\xi)$. Then for every $((\bar{a}, \mu, \sigma), \bar{t}) \in T((\Sigma^{(n)})_{\mathcal{V}})^*$ with $((\bar{a}, \sigma), \bar{t}) \in N_{\mathcal{V}}$ we have*

$$((\bar{a}, \sigma), \bar{t}) \models \xi \quad \textit{if and only if} \quad ((\bar{a}, \mu, \sigma), \bar{t}) \models \widetilde{\xi}.$$

First-Order Universal Quantification Let $\psi \in \mathsf{MSO}(\mathcal{K}, T\Sigma^*, \mathcal{F})$ be almost unambiguous over y and assume $\varphi = \forall y.\psi$. By Lemma 5.12, we may assume that ψ is of the form

$$\psi = \bigvee_{1 \leq j \leq n} f_j(y) \wedge k_j \wedge \psi_j$$

where $n \in \mathbb{N}$, $k_j \in K$, $f_j \in \mathcal{F}$, unweighted $\psi_j \in \mathsf{MSO}(\mathcal{K}, T\Sigma^*, \mathcal{F})$ for each $j \in \{1, ..., n\}$.

Let $\mathcal{W} = \mathsf{Free}(\psi)$ and $\mathcal{V} = \mathsf{Free}(\varphi) = \mathcal{W}\setminus\{y\}$. Recall that $\psi_1, ..., \psi_n$ can be considered as formulas in $\mathsf{MSO}(T\Sigma^*)$. We may assume that $\psi_1, ..., \psi_n$ define a partition of $N_{\mathcal{W}}$. We define $\widetilde{L} \subseteq T((\Sigma^{(n)})_{\mathcal{V}})^*$ to be the set of timed words $((\bar{a}, \mu, \sigma), \bar{t})$ in $T((\Sigma^{(n)})_{\mathcal{V}})^*$ such that $((\bar{a}, \sigma), \bar{t}) \in N_{\mathcal{V}}$, and for all $i \in \mathsf{dom}(\bar{a}, \bar{t})$ and $j \in \{1, ..., n\}$ we have

$$\mu(i) = j \text{ implies } ((\bar{a}, \sigma[y \to i]), \bar{t}) \models \psi_j.$$

Notice that for every $((\bar{a}, \sigma), \bar{t}) \in N_{\mathcal{V}}$, there is a unique μ such that $((\bar{a}, \mu, \sigma), \bar{t}) \in \widetilde{L}$, since $(\psi_1, ..., \psi_n)$ forms a partition of $N_{\mathcal{W}}$. Next, we prove that $\widetilde{L}$ is TA-recognizable. For this, consider the formula $\zeta \in \mathsf{MSO}(T(\Sigma^{(n)})^*)$

$$\zeta = \forall y. \bigwedge_{1 \leq j \leq n} \bigwedge_{a \in \Sigma} \left(P_{(a,j)}(y) \longrightarrow \widetilde{\psi_j} \right).$$

Let $((\bar{a}, \mu, \sigma), \bar{t}) \in T((\Sigma^{(n)})_{\mathcal{V}})^*$ such that $((\bar{a}, \sigma), \bar{t}) \in N_{\mathcal{V}}$. Using the semantics of $\mathsf{MSO}(T(\Sigma^{(n)})^*)$, one can show that $((\bar{a}, \mu, \sigma), \bar{t}) \models \zeta$ if and only if for every $i \in \mathsf{dom}(\bar{a}, \bar{t})$ and $j \in \{1, ..., n\}$ we have that $\mu(i) = j$ implies $((\bar{a}, \mu, \sigma[y \to i]), \bar{t}) \models \widetilde{\psi_j}$. This, by Lemma 5.16, holds if and only if $((\bar{a}, \sigma[y \to i]), \bar{t}) \models \psi_j$. Thus, $((\bar{a}, \mu, \sigma), \bar{t}) \models \zeta$ if and only if $((\bar{a}, \mu, \sigma), \bar{t}) \in \widetilde{L}$, and we have $L(\zeta) = \widetilde{L}$. By Theorem 5.2, $\widetilde{L}$ is TA-recognizable over $(\Sigma^{(n)})_{\mathcal{V}}$.

Next, we will use the information encoded in μ to build a weighted timed automaton over $\mathcal{K}$, $(\Sigma^{(n)})_{\mathcal{V}}$ and $\mathcal{F}$. Let $\widetilde{\mathcal{A}} = (\mathcal{L}, \mathcal{L}_0, \mathcal{L}_f, \mathcal{C}, E)$ be a timed automaton such that $L(\widetilde{\mathcal{A}}) = \widetilde{L}$. By Lemma 5.15, there is a timed automaton $\mathcal{A}' = (\mathcal{L}', \mathcal{L}'_0, \mathcal{L}'_f, \mathcal{C}, E')$ such that $L(\mathcal{A}') = L(\widetilde{\mathcal{A}})$, the locations in $\mathcal{A}'$ are elements in $\mathcal{L} \times (\Sigma^{(n)})_{\mathcal{V}}$, and for each $(l, (a, b, \sigma)) \in \mathcal{L}'$, every outgoing edge is labeled with (a, b, σ). Observe that this latter fact is crucial for assigning the weight functions to the locations in a proper way. Now define $\mathcal{A} = (\mathcal{L}', \mathcal{C}, E', \mathsf{in}, \mathsf{out}, \mathsf{ewt}, \mathsf{lwt})$ by

- $\mathsf{in}(l) = 1$ if $l \in \mathcal{L}'_0$, and $\mathsf{in}(l) = 0$ otherwise,
- $\mathsf{out}(l) = 1$ if $l \in \mathcal{L}'_f$, and $\mathsf{out}(l) = 0$ otherwise,
- $\mathsf{ewt}\big((l, (a, b, \sigma)), (a, b, \sigma), \phi, \lambda, (l', (a', b', \sigma'))\big) = k_b$ for each $\big((l, (a, b, \sigma)), (a, b, \sigma), \phi, \lambda, (l', (a', b', \sigma'))\big) \in E'$,
- $\mathsf{lwt}\big((l, (a, b, \sigma))\big) = f_b$ for every $(l, (a, b, \sigma)) \in \mathcal{L}'$.

Note that $\mathsf{Func}(\mathcal{A}) = \{f_1, ..., f_n\}$. We also observe that for each $w = ((\bar{a}, \mu, \sigma), \bar{t}) \in T((\Sigma^{(n)})_{\mathcal{V}})^*$, and for each run r of $\mathcal{A}$ on w with $\mathsf{rwt}(r) \neq 0$ we have

$$\mathsf{rwt}(r) = \prod_{i \in \mathsf{dom}(\bar{a}, \bar{t})} f_{\mu(i)}(t_i - t_{i-1}) \cdot k_{\mu(i)}. \tag{5.1}$$

Consider the renaming $p : (\Sigma^{(n)})_{\mathcal{V}} \to \Sigma_{\mathcal{V}}$ defined by $(a, b, \sigma) \mapsto (a, \sigma)$ for each $(a, b, \sigma) \in (\Sigma^{(n)})_{\mathcal{V}}$. We show that $\bar{p}(\|\mathcal{A}\|) = [\![\forall y.\psi]\!]$. First, for every $((\bar{a}, \sigma), \bar{t}) \in N_{\mathcal{V}}$ and the unique μ such that $((\bar{a}, \mu, \sigma), \bar{t}) \in \widetilde{L}$, we have

$$\begin{aligned}
\big(\bar{p}(\|\mathcal{A}\|), ((\bar{a}, \sigma), \bar{t})\big) &= \big(\|\mathcal{A}\|, ((\bar{a}, \mu, \sigma), \bar{t})\big) \\
&\stackrel{\star}{=} \prod_{i \in \mathsf{dom}(\bar{a}, \bar{t})} f_{\mu(i)}(t_i - t_{i-1}) \cdot k_{\mu(i)} \\
&= \prod_{i \in \mathsf{dom}(\bar{a}, \bar{t})} \big([\![\varphi]\!]_{\mathcal{W}}, ((\bar{a}, \sigma[y \to i]), \bar{t})\big) \\
&= \big([\![\forall y.\varphi]\!], ((\bar{a}, \sigma), \bar{t})\big)
\end{aligned}$$

where $\star$ is due to (1) and idempotence of $\mathcal{K}$. For every $((\bar{a}, \sigma), \bar{t}) \notin N_{\mathcal{V}}$, we obtain 0 for both $\big(\bar{p}(\|\mathcal{A}\|), ((\bar{a}, \sigma), \bar{t})\big)$ and $\big([\![\forall y.\varphi]\!], ((\bar{a}, \sigma), \bar{t})\big)$. Thus, $\bar{p}(\mathcal{A}) = [\![\forall y.\psi]\!]$.

Finally, let $\mathcal{A}_\varphi$ be the weighted timed automaton over $\Sigma_{\mathcal{V}}$ obtained from $\mathcal{A}$ as defined in the proof of Lemma 3.7. Hence, we have $\|\mathcal{A}_\varphi\| = \bar{p}(\|\mathcal{A}\|) = [\![\varphi]\!]$. Since the construction of $\mathcal{A}_\varphi$ according to Lemma 3.7 does not add any location weight functions, condition 2. is satisfied. Condition 3 is trivially satisfied, since the set of subformulas $f(z)$ occurring in φ with $z \in \mathsf{Free}(\varphi)$ is empty. This finishes the proof of Theorem 5.9.

■

We proved that each formula $\varphi \in \mathsf{sRMSO}(\mathcal{K}, T\Sigma^*, \mathcal{F})$ is recognizable by a weighted timed automaton. Now, we give the definition of the **syntactically restricted weighted relative distance logic**, denoted by $\mathsf{sR}\overleftarrow{\mathcal{L}\mathsf{d}}(\mathcal{K}, \Sigma, \mathcal{F})$. It is defined as the smallest class of formulas containing all formulas generated by the next two rules.

1. If $\varphi \in \mathsf{sRMSO}(\mathcal{K}, T\Sigma^*, \mathcal{F})$, then $\varphi \in \mathsf{sR}\mathcal{L}\overleftarrow{\mathsf{d}}\,(\mathcal{K}, \Sigma, \mathcal{F})$.
2. If $\varphi \in \mathsf{sR}\mathcal{L}\overleftarrow{\mathsf{d}}\,(\mathcal{K}, \Sigma, \mathcal{F})$, then $\exists D.\varphi \in \mathsf{sR}\mathcal{L}\overleftarrow{\mathsf{d}}\,(\mathcal{K}, \Sigma, \mathcal{F})$.

Altogether, using the same lines of argumentation as in the proof of Theorem 5.9 in the case of existential quantification, we can show that if the semantics of $\varphi \in \mathsf{sR}\mathcal{L}\overleftarrow{\mathsf{d}}\,(\mathcal{K}, \Sigma, \mathcal{F})$ is $\mathcal{F}$-recognizable over $\Sigma_{\mathsf{Free}(\varphi)}$, so is the semantics of $\exists D.\varphi$ $\mathcal{F}$-recognizable over $\Sigma_{\mathsf{Free}(\exists D.\varphi)}$. Altogether, we obtain the following theorem, which corresponds to one direction of a Büchi theorem for the class of $\mathcal{F}$-recognizable timed series.

Theorem 5.17. *Let $\mathcal{K}$ be idempotent and commutative. If $\varphi \in \mathsf{sR}\mathcal{L}\overleftarrow{\mathsf{d}}\,(\mathcal{K}, \Sigma, \mathcal{F})$, then $[\![\varphi]\!]$ is $\mathcal{F}$-recognizable over $\Sigma_{\mathsf{Free}(\varphi)}$.*

5.4 From Recognizability to Definability

In this section, we show that the behaviour of each weighted timed automaton over $\mathcal{K}$, Σ and $\mathcal{F}$ can be defined by a sentence in $\mathsf{sR}\mathcal{L}\overleftarrow{\mathsf{d}}\,(\mathcal{K}, \Sigma, \mathcal{F})$. For this, we extend the proof developed by Droste and Gastin for series to the timed series.

Theorem 5.18. *Let $\mathcal{K}$ be idempotent and commutative. Each $\mathcal{F}$-recognizable timed series is $\mathsf{sR}\mathcal{L}\overleftarrow{\mathsf{d}}\,(\mathcal{K}, \Sigma, \mathcal{F})$-definable.*

Proof. Let $\mathcal{T} : T\Sigma^* \to K$ be $\mathcal{F}$-recognizable. Then there is a weighted timed automaton $\mathcal{A} = (\mathcal{L}, \mathcal{C}, E, \mathsf{in}, \mathsf{out}, \mathsf{ewt}, \mathsf{lwt})$ such that $\|\mathcal{A}\| = \mathcal{T}$. We choose an enumeration $(x_1, ..., x_m)$ of $\mathcal{C}$ together with an enumeration $(e_1, ..., e_n)$ of E and assume $e_i = (l_i, a_i, \phi_i, \lambda_i, l_i')$. Let $\bar{D} = D_1, ..., D_m$, where D_i stands for the clock variable x_i for each $i \in \{1, ..., m\}$, and let $\bar{Y} = Y_1, ..., Y_n$, where Y_j stands for the edge e_j for each $j \in \{1, ..., n\}$. We define an unweighted formula $\psi(\bar{D}, \bar{Y}) \in \mathsf{MSO}(\mathcal{K}, T\Sigma^*, \mathcal{F})$ describing the runs of $\mathcal{A}$.

First, we define

$$\psi_{\text{partition}} := \forall y. \bigvee_{1 \leq i \leq n} \Big(y \in Y_i \wedge \bigwedge_{\substack{1 \leq j \leq n \\ i \neq j}} \neg(y \in Y_j)\Big)$$

$$\psi_{\text{label}} := \forall y. \bigwedge_{a \in \Sigma} \Big(P_a(y) \longrightarrow \big(\bigvee_{\substack{1 \leq i \leq n \\ a_i = a}} y \in Y_i\big)\Big)$$

$$\psi_{\text{consistent}} := \forall y. \forall z. (z = y + 1) \longrightarrow \bigvee_{\substack{1 \leq i,j \leq n: \\ l_i' = l_j}} (y \in Y_i \wedge z \in Y_j)$$

$$\psi_{\text{test}} := \forall y. \bigwedge_{1 \leq i \leq n} \left(y \in Y_i \longrightarrow \bigwedge_{(x_j \sim c) \in \phi_i} \overleftarrow{\mathsf{d}}\,(D_j, y) \sim c \right)$$

$$\psi_{\text{reset}} \quad := \quad \bigwedge_{1 \leq i \leq m} \forall y. y \in D_i \longleftrightarrow \bigvee_{\substack{1 \leq j \leq n \\ x_i \in \lambda_j}} y \in Y_j$$

where $y \leq z := y < z \vee y = z$, $(z = y+1) := y \leq z \wedge \neg(z \leq y) \wedge \forall z'.\big((z' \leq y) \vee (z \leq z')\big)$. Finally, put $\psi(\bar{D}, \bar{Y}) = \psi_{\text{partition}} \wedge \psi_{\text{label}} \wedge \psi_{\text{consistent}} \wedge \psi_{\text{test}} \wedge \psi_{\text{reset}}$.

By Lemma 5.10, $[\![\psi(\bar{D}, \bar{Y})]\!] = 1_{L(\psi(\bar{D},\bar{Y}))}$. We define $\mathcal{V} = \{D_1, ..., D_m, Y_1, ..., Y_n\}$ and observe that $\mathsf{Free}(\psi(\bar{D}, \bar{Y})) = \mathcal{V}$. Let $w = (\bar{a}, \bar{t}) \in T\Sigma^*$. We show that there is a bijective correspondence between the set of runs of $\mathcal{A}$ on w and the set of $(\mathcal{V}, w)$-assignments σ with $([\![\psi(\bar{D}, \bar{Y})]\!], ((\bar{a}, \sigma), \bar{t})) = 1$.

Construction 1 Let $r = (l_0, \nu_0) \xrightarrow{\delta_1} \xrightarrow{e_1} \dots \xrightarrow{\delta_{|w|}} \xrightarrow{e_{|w|}} (l_{|w|}, \nu_{|w|})$ be a run of $\mathcal{A}$ on w. We define the $(\mathcal{V}, w)$-assignment σ_r by $\sigma_r(D_j) = \{i : x_j \in \lambda_i\}$ and $\sigma_r(Y_j) = \{i : e_i = e_j\}$. Intuitively, $\sigma_r(D_j)$ contains exactly the positions $i \in \mathsf{dom}(w)$, where the clock x_j is reset, and $\sigma_r(Y_j)$ contains exactly the positions $i \in \mathsf{dom}(w)$ that arose after the edge e_j has been executed. Then we have $\big([\![\psi(\bar{D}, \bar{Y})]\!], ((\bar{a}, \sigma_r), \bar{t})\big) = 1$.

Construction 2 Let σ be a valid $(\mathcal{V}, w)$-assignment such that $\big([\![\psi(\bar{D}, \bar{Y})]\!], ((\bar{a}, \sigma), \bar{t})\big) = 1$. Using σ, we construct a unique run $r_\sigma = (l_0, \nu_0) \xrightarrow{\delta_1} \xrightarrow{e_1} \dots \xrightarrow{\delta_{|w|}} \xrightarrow{e_{|w|}} (l_{|w|}, \nu_{|w|})$ of $\mathcal{A}$ on w as follows:

- $\delta_i = t_i - t_{i-1}$ for each $i \in \mathsf{dom}(w)$,
- $e_i = e_j$ such that $i \in \sigma(Y_j)$ for each $i \in \mathsf{dom}(w)$. Note that due to $\psi_{\text{partition}}$, there is exactly one such Y_j.

It can be seen very easily, that r_σ is a run of $\mathcal{A}$ on w. This establishes the bijective correspondence mentioned above. Then, for every valid $(\mathcal{V}, w)$-assignment σ, we have $\big([\![\psi(\bar{D}, \bar{Y})]\!], ((\bar{a}, \sigma), \bar{t})\big) = 1$ if there is a run of $\mathcal{A}$ on w, and $\big([\![\psi(\bar{D}, \bar{Y})]\!], ((\bar{a}, \sigma), \bar{t})\big) = 0$ otherwise.

Now, we "add weights" to $\psi(\bar{D}, \bar{Y})$ to obtain a formula $\xi(\bar{D}, \bar{Y})$ whose semantics corresponds to the running weight of a run of $\mathcal{A}$ on $(\bar{a}, \bar{t})$. Define

$$\begin{aligned}
\xi(\bar{D}, \bar{Y}) = \psi(\bar{D}, \bar{Y}) \quad & \wedge \quad \exists y. \big(\forall z. y \leq z \wedge \bigvee_{e_i \in E} y \in Y_i \wedge \mathsf{in}(l_i)\big) \\
& \wedge \quad \bigwedge_{e_i \in E} \forall y. \big(\neg(y \in Y_i) \vee [y \in Y_i \wedge \mathsf{lwt}(l_i)(y) \wedge \mathsf{ewt}(e_i)]\big) \\
& \wedge \quad \exists y. \big(\forall z. z \leq y \wedge \bigvee_{e_i \in E} y \in Y_i \wedge \mathsf{out}(l_i')\big).
\end{aligned}$$

Now, in general, for every $k \in K$, $f \in \mathcal{F}$ and valid $(\mathcal{V}, w)$-assignment, we have

$$\big(\llbracket \neg(y \in X) \vee (y \in X \wedge f(y) \wedge k)) \rrbracket_{\mathcal{V}}, ((\bar{a}, \sigma), \bar{t})\big) = \begin{cases} f(t_{\sigma(y)} - t_{\sigma(y)-1}) \cdot k & \text{if } \sigma(y) \in \sigma(X) \\ 1 & \text{otherwise} \end{cases}$$

and thus

$$\big(\llbracket \neg(y \in X) \vee (y \in X \wedge f(y) \wedge k)) \rrbracket_{\mathcal{V}}, ((\bar{a}, \sigma), \bar{t})\big) = (f(t_{\sigma(y)} - t_{\sigma(y)-1}) \cdot k)^{|\sigma(X)|}.$$

This will be used in the following. Let $r = (l_0, \nu_0) \xrightarrow{\delta_1} \xrightarrow{e_1} \ldots \xrightarrow{\delta_2} \xrightarrow{e_{|w|}} (l_{|w|}, \nu_{|w|})$ be a run of $\mathcal{A}$ on w. Further, we let σ_r be the associated $(\mathcal{V}, w)$-assignment (see construction 1 above). Then, we have

$$\begin{aligned} & \big(\llbracket \xi(\bar{D}, \bar{Y}) \rrbracket, ((\bar{a}, \sigma_r), \bar{t})\big) \\ = \;& \mathsf{in}(l_0) \cdot \left(\prod_{1 \leq i \leq |w|} \big(\mathsf{lwt}(l_i)(t_{\sigma(y)} - t_{\sigma(y)-1}) \cdot \mathsf{ewt}(e_i)\big)^{|\sigma_r(Y_i)|} \right) \cdot \mathsf{out}(l_{|w|}) \\ = \;& \mathsf{rwt}(r). \end{aligned}$$

Notice that due to the subformulas starting with existential first-order quantification, we have $\big(\llbracket \psi(\bar{D}, \bar{Y}) \rrbracket, \varepsilon\big) = 0$. Thus, we need to construct a sentence that is equivalent to the behaviour of $\mathcal{A}$ for ε. This can be done as in the classical case, i.e., for instance we can choose $\varphi = (\|\mathcal{A}\|, \varepsilon) \wedge \forall y. \neg(y \leq y)$. For $w \in \Sigma^+$, we obtain $\big(\llbracket \forall y. \neg(y \leq y) \rrbracket, w\big) = 0$, whereas $\big(\llbracket \forall y. \neg(y \leq y) \rrbracket, \varepsilon\big) = 1$ since an empty product is 1 by convention. Hence, $\big(\llbracket \varphi \rrbracket, \varepsilon\big) = (\|\mathcal{A}\|, \varepsilon)$. Now, define $\zeta = \exists D_1 \ldots \exists D_m \exists Y_1 \ldots \exists Y_n . \xi(\bar{D}, \bar{Y})$. Observe that $\zeta \in \mathsf{sR}\overleftarrow{\mathcal{L}\mathsf{d}}(\mathcal{K}, \Sigma, \mathcal{F})$. We further have $\big(\llbracket \zeta \rrbracket, \varepsilon\big) = \big(\llbracket \varphi \rrbracket, \varepsilon\big) = (\|\mathcal{A}\|, \varepsilon)$. Using the bijective correspondence from above, we obtain for every $(\bar{a}, \bar{t}) \in T\Sigma^+$

$$\begin{aligned} \big(\llbracket \zeta \rrbracket, ((\bar{a}, \bar{t}))\big) &= \sum_{\substack{\sigma \\ (\mathcal{V}, (\bar{a}, \bar{t}))-\text{assignment}}} \big(\llbracket \xi \rrbracket, ((\bar{a}, \sigma), \bar{t})\big) \\ &= \sum_{\substack{r \\ \text{run of } \mathcal{A}}} \big(\llbracket \xi \rrbracket, ((\bar{a}, \sigma_r), \bar{t})\big) \\ &= \sum_{\substack{r \\ \text{run of } \mathcal{A}}} \mathsf{rwt}(r) \\ &= (\|\mathcal{A}\|, (\bar{a}, \bar{t})). \end{aligned}$$

■

As a consequence of Theorems 5.17 and 5.18, we obtain a Büchi theorem for the class of $\mathcal{F}$-recognizable timed series over idempotent and commutative semirings.

Theorem 5.19. *Let $\mathcal{K}$ be idempotent and commutative and $\mathcal{F}$ contain $\mathbb{1}$. Each timed series $\mathcal{T} : T\Sigma^* \to K$ is $\mathcal{F}$-recognizable if and only if $\mathcal{T}$ is $\overleftarrow{\mathsf{sR}\mathcal{L}\mathsf{d}}\,(\mathcal{K}, \Sigma, \mathcal{F})$-definable.*

We remark that the respective transformations can be done effectively provided that the operations of $\mathcal{K}$ and $\mathcal{F}$ are given effectively. For the direction from weighted timed automata to sentences in $\overleftarrow{\mathsf{sR}\mathcal{L}\mathsf{d}}\,(\mathcal{K}, \Sigma, \mathcal{F})$ this claim is obvious. For the other direction, we point out that the only critical point in the proof is the construction of a weighted timed automaton recognizing $[\![\forall y.\varphi]\!]$ if φ is almost unambiguous. However, by Lemma 5.12 we can transform each almost unambiguous formula into the form $\bigvee_{1 \leq i \leq n} f_i(y) \wedge k_i \wedge \psi_i$ with $f_i \in \mathcal{F}$, $k_i \in K$ and unweighted $\psi_i \in \mathsf{MSO}(\mathcal{K}, T\Sigma^*, \mathcal{F})$ for each $i \in \{1, ..., n\}$ and some $n \in \mathbb{N}$ as it is required in the corresponding construction.

5.5 Generalizations to Arbitrary Semirings

In this section, we explain how we can generalize Theorem 5.19 to non-idempotent semirings. Later on, we will indicate how we skip the restriction on the semiring being commutative.

In the following, let $\mathcal{K}$ be a commutative semiring, not necessarily being idempotent. In the last section, we used the idempotence of $\mathcal{K}$ in two crucial steps. First, in Lemma 5.10, where we claimed that each unweighted formula $\psi \in \mathsf{MSO}(\mathcal{K}, T\Sigma^*, \mathcal{F})$ takes only values in $\{0, 1\}$. This no longer holds if $\mathcal{K}$ is not idempotent. We thus cannot use Lemma 5.10 to show Lemmas 5.11 and 5.12. Notice that if we exclude the usage of disjunction and existential quantification in an unweighted formula $\varphi \in \mathsf{MSO}(\mathcal{K}, T\Sigma^*, \mathcal{F})$, then the semantics of φ takes only values in $\{0, 1\}$. However, not every unweighted formula can be transformed into a language equivalent such formula, due to the syntactical restriction on negation. Instead, we will introduce *syntactically unambiguous formulas.*

Second, we used the idempotence of $\mathcal{K}$ in the proof of universal first-order quantification. We used that for each timed word w, the running weights of all runs of $\mathcal{A}$ on w are the same, and thus, by idempotence of $\mathcal{K}$, the behaviour of $\mathcal{A}$ on w is the same as the running weight of an arbitrary run of $\mathcal{A}$ on w. Notice that if there is exactly one run of $\mathcal{A}$ on w, i.e., $\mathcal{A}$ is unambiguous, then the behaviour of $\mathcal{A}$ on w is the same as the running weight of this run. However, as we have noted in Sect. 2.1, the class of unambiguously TA-recognizable timed languages is a strict subclass of TA-recognizable timed languages. For this reason, we focus on a subclass of timed languages whose elements have a *bounded variability.* Then we take advantage of the fact that every such timed language is *deterministically* (and thus, unambiguously) TA-recognizable.

Bounded Variability of Timed Languages The notion of bounded variability of timed words has been introduced by Wilke [107]. Intuitively, the variability of a timed word corresponds to the maximum number of events that may occur within one time unit. When bounding the variability of timed words, we can always construct deterministic timed automata. Using this, Wilke showed that $\mathcal{L}\overleftarrow{\mathsf{d}}(\Sigma)$ is fully decidable over the class of timed languages with bounded variability (as opposed to the class of all timed languages) [107]. The restriction to timed languages with bounded variability is a reasonable assumption as practically any system can only handle a bounded number of tasks within a time unit. Recently, another positive decidability result concerning MTL model checking was shown for this particular class of timed languages [66].

Let $M \subseteq T\Sigma^*$ be a set of timed words. We say that $L \subseteq T\Sigma^*$ is *TA-recognizable over* Σ *relatively to* M if there is a timed automaton $\mathcal{A}$ over Σ such that $L = L(\mathcal{A}) \cap M$. Let $w = (a_1, t_1)...(a_k, t_k) \in T\Sigma^*$. The *variability* of w, denoted by $var(w)$, is defined as $\sup\{b + 1 : \exists i.1 \leq i \leq k - b \text{ and } t_{i+b} - t_i < 1\}$. Intuitively, the variability of a timed word gives the maximum number of events in a unit time interval. We say that w is of bounded variability b for some $b \in \mathbb{N}$ if the variability of w is less than or equal to b. We use $T_b\Sigma^*$ to denote the set $\{w \in T\Sigma^* : var(w) \leq b\}$ of all timed words of bounded variability b. By bounding the variability of a timed word we fix the maximum number of events in a unit time interval.

Remark 5.20. In the literature, there are also other restrictions on the occurence of events within timed words, the most known of which is the restriction of being *non-Zeno*. A timed word is non-Zeno if the sequence of timestamps of the word is diverging. Hence, every finite word is non-Zeno and thus this notion is weaker than that of bounded variability. The restriction of being non-Berkeley for some positive real number δ has been introduced by Furia and Rossi [66] and means that between any two events more than δ time units must pass. For a comparison between these three restrictions see the paper of Furia and Rossi.

In the following, we fix a bound $b \in \mathbb{N}$.

Proposition 5.21 ([106]).

1. *If $L \subseteq T\Sigma^*$ is TA-recognizable over Σ, we can effectively construct a deterministic timed automaton $\mathcal{A}$ over Σ such that $L(\mathcal{A}) = L \cap T_b\Sigma^*$.*
2. *The class of TA-recognizable timed languages over Σ relatively to $T_b\Sigma^*$ is closed under boolean operations, renamings and inverse renamings.*
3. *The set $T_b\Sigma^*$ is $\mathcal{L}\overleftarrow{\mathsf{d}}(\Sigma)$-definable.*

We let $\exists D_1...\exists D_b.\varphi_b$ denote a sentence in $\mathcal{L}\overleftarrow{\mathsf{d}}(\Sigma)$ defining $T_b\Sigma^*$. For instance, φ_b may be the formula

$$\varphi_b = \begin{pmatrix} (1 \in D_1 \wedge 2 \in D_2 \wedge ... \wedge b \in D_b) \\ \wedge \\ \bigwedge_{1 \leq i \leq b} \forall y.(y \in D_i \longleftrightarrow (y + b) \in D_i) \\ \wedge \\ \bigwedge_{1 \leq i \leq b} \forall y.(y \in D_i \longrightarrow \overleftarrow{\mathsf{d}}(D_i, y) < 1) \end{pmatrix}$$

where $1, 2, ..., b$ stand for the first, second,..., b-th position in w, and $(y + b)$ stands for the b-th position in w after y. These terms can easily expressed in $\mathsf{MSO}(T\Sigma^*)$.

Syntactically Unambiguous Formulas We define for each unweighted formula $\varphi \in \mathsf{MSO}(\mathcal{K}, T\Sigma^*, \mathcal{F})$ a language equivalent formula ψ that has at most one assignment evaluating a timed word to 1. Since in general there may be more than one such assignment, we choose the *first* such assignment, in the following sense: if y is a free first-order variable, then we choose the smallest position with a one in the y-row; if X is a free second-order variable, then we choose the set of the smallest positions with a one in the X-row.

Let $\varphi, \xi \in \mathsf{MSO}(\mathcal{K}, T\Sigma^*, \mathcal{F})$ be unweighted. We define the formulas φ^+, φ^-, $\varphi \xrightarrow{+} \xi$ and $\varphi \xleftrightarrow{+} \xi$ inductively as follows.

1. If φ is of the form $P_a(y)$, $y < z$, $y = z$, $y \in X$, $\overleftarrow{\mathsf{d}}(D, y) \sim c$, then $\varphi^+ = \varphi$ and $\varphi^- = \neg\varphi$.
2. If $\varphi = \neg\psi$, then $\varphi^+ = \psi^-$ and $\varphi^- = \psi^+$.
3. If $\varphi = \psi \vee \zeta$, then $\varphi^+ = \psi^+ \vee (\psi^- \wedge \zeta^+)$ and $\varphi^- = \psi^- \wedge \zeta^-$.
4. If $\varphi = \psi \wedge \zeta$, then $\varphi^- = \psi^- \vee (\psi^+ \wedge \zeta^-)$ and $\varphi^+ = \psi^+ \wedge \zeta^+$.
5. If $\varphi = \exists y.\psi$, then $\varphi^+ = \exists y.(\psi^+(y) \wedge \forall z.(z < y \wedge \psi(z))^-)$ and $\varphi^- = \forall y.\psi^-$.
6. If $\varphi = \forall y.\psi$, then $\varphi^- = \exists y.(\psi^-(y) \wedge \forall z.(y \leq z \vee \psi(z))^+)$ and $\varphi^+ = \forall y.\psi^+$.
7. $\varphi \xrightarrow{+} \xi = \varphi^- \vee (\varphi^+ \wedge \xi^+)$
8. $\varphi \xleftrightarrow{+} \xi = (\varphi^+ \wedge \xi^+) \vee (\varphi^- \wedge \xi^-)$
9. For second-order variables X, Y, we define

$$\begin{aligned} X = Y &= \forall y.(y \in X \xleftrightarrow{+} y \in Y), \\ X < Y &= \exists y.(y \in Y \wedge \neg(y \in X) \wedge \forall z.[z < y \xrightarrow{+} (z \in X \xleftrightarrow{+} z \in Y)]), \\ X \leq Y &= (X = Y) \vee (X < Y). \end{aligned}$$

10. If $\varphi = \exists X.\psi$, then $\varphi^+ = \exists X.(\psi^+(X) \wedge \forall Y.(Y < X \wedge \psi(Y))^-)$ and $\varphi^- = \forall X.\psi^-$.
11. If $\varphi = \forall X.\psi$, then $\varphi^- = \exists X.(\psi^-(X) \wedge \forall Y.(X \leq Y \vee \psi(Y))^+)$ and $\varphi^+ = \forall X.\psi^+$.

We define the class of *syntactically unambiguous* formulas in $\mathsf{MSO}(\mathcal{K}, T\Sigma^*, \mathcal{F})$ as the smallest class of formulas containing all formulas of the form

- $\varphi^+, \varphi^-, \varphi \xrightarrow{+} \xi$ and $\varphi \xleftrightarrow{+} \xi$ if $\varphi, \xi \in \mathsf{MSO}(\mathcal{K}, T\Sigma^*, \mathcal{F})$ are unweighted, and

- $\forall y.\varphi, \forall X.\varphi$ or $\varphi \wedge \psi$ if it contains φ and ψ.

We say that a formula $\varphi \in \mathsf{MSO}(\mathcal{K}, T\Sigma^*, \mathcal{F})$ is *syntactically unambiguous of bounded variability* b if it is of the form $\psi \wedge (\varphi_b)^+$ for some syntactically unambiguous formula ψ. Similarly, we say that φ is *almost unambiguous over* y *of bounded variability* b if it is in the disjunctive and conjunctive closure of syntactically unambiguous formulas of bounded variability b, constants $k \in K$ and formulas $f(y)$ for some $f \in \mathcal{F}$, such that $f(y)$ may appear at most once in every subformula of φ of the form $\varphi_1 \wedge \varphi_2$.

We define the **syntactically restricted auxiliary logic of bounded variability** b $\mathsf{sRMSO}^{\mathsf{b}}(\mathcal{K}, T\Sigma^*, \mathcal{F})$ to be the smallest class of formulas generated by the following rules.

1. If $\varphi \in \mathsf{MSO}(\mathcal{K}, T\Sigma^*, \mathcal{F})$ is an atomic or a weighted atomic formula, then $\varphi \in \mathsf{sRMSO}^{\mathsf{b}}(\mathcal{K}, T\Sigma^*, \mathcal{F})$.
2. If $\varphi, \psi \in \mathsf{sRMSO}^{\mathsf{b}}(\mathcal{K}, T\Sigma^*, \mathcal{F})$, then $\varphi \vee \psi, \exists y.\varphi, \exists X.\varphi \in \mathsf{sRMSO}^{\mathsf{b}}(\mathcal{K}, T\Sigma^*, \mathcal{F})$.
3. If $\varphi \in \mathsf{sRMSO}^{\mathsf{b}}(\mathcal{K}, T\Sigma^*, \mathcal{F})$ is syntactically unambiguous of bounded variability b, then $\forall X.\varphi \in \mathsf{sRMSO}^{\mathsf{b}}(\mathcal{K}, T\Sigma^*, \mathcal{F})$.
4. If $\varphi \in \mathsf{sRMSO}^{\mathsf{b}}(\mathcal{K}, T\Sigma^*, \mathcal{F})$ is almost unambiguous over y of bounded variability b, then $\forall y.\varphi \in \mathsf{sRMSO}^{\mathsf{b}}(\mathcal{K}, T\Sigma^*, \mathcal{F})$.
5. If $\varphi, \psi \in \mathsf{sRMSO}^{\mathsf{b}}(\mathcal{K}, T\Sigma^*, \mathcal{F})$ and at least one of the following conditions holds
   - $\mathcal{V}_f(\varphi) = \emptyset$,
   - $\mathcal{V}_f(\psi) = \emptyset$,
   - $\mathcal{V}_f(\varphi) \cap \mathcal{V}_f(\psi) = \emptyset$, $\mathcal{V}_f(\varphi) \subseteq \mathsf{Free}(\varphi)$, and $\mathcal{V}_f(\psi) \subseteq \mathsf{Free}(\psi)$,

   then $\varphi \wedge \psi \wedge \bigwedge_{\substack{y \in \mathcal{V}_f(\varphi), z \in \mathcal{V}_f(\psi) \\ y \neq z}} \neg(y = z) \in \mathsf{sRMSO}^{\mathsf{b}}(\mathcal{K}, T\Sigma^*, \mathcal{F})$.

Note that the definition of $\mathsf{sRMSO}^{\mathsf{b}}(\mathcal{K}, T\Sigma^*, \mathcal{F})$ differs from the definition of $\mathsf{sRMSO}(\mathcal{K}, T\Sigma^*, \mathcal{F})$ only in rules 3. and 4. Next, we want to prove the following theorem.

Theorem 5.22. *Let* $\varphi \in \mathsf{sRMSO}^{\mathsf{b}}(\mathcal{K}, T\Sigma^*, \mathcal{F})$. *Then for each finite set* $\mathcal{V} \supseteq \mathsf{Free}(\varphi)$ *there is some weighted timed automaton* $\mathcal{A}_\varphi$ *over* $\mathcal{K}$, $\Sigma_\mathcal{V}$ *and* $\mathcal{F}$ *such that*

1. $\|\mathcal{A}_\varphi\| = [\![\varphi]\!]_\mathcal{V}$,
2. $\mathsf{Func}(\mathcal{A}_\varphi) \subseteq \mathsf{Func}(\varphi) \cup \{\mathbb{1}\}$,
3. *for each formula* $f(y)$ *occurring in* φ *with* $y \in \mathsf{Free}(\varphi)$, *whenever* $\mathsf{lwt}(l) = f$ *for some location* l *in* $\mathcal{A}_\varphi$, *then for each edge* $(l, (a, \sigma), \phi, \lambda, l')$ *in* $\mathcal{A}_\varphi$ *we have* $\sigma(y) = 1$.

The proof is along the lines of the proof of Theorem 5.9. However, we have to give new proofs for both universal quantifiers. We start with some lemmas. By induction it is easy to show:

Lemma 5.23. *Let $\varphi \in \mathsf{MSO}(\mathcal{K}, T\Sigma^*, \mathcal{F})$ be unweighted. Then we have*

1. $L(\varphi^+) = L(\varphi)$ *and* $L(\varphi^-) = L(\neg\varphi)$,
2. $[\![\varphi^+]\!] = 1_{L(\varphi)}$ *and* $[\![\varphi^-]\!] = 1_{L(\neg\varphi)}$.

Lemma 5.24. *Let $\psi_1, \psi_2 \in \mathsf{MSO}(\mathcal{K}, T\Sigma^*, \mathcal{F})$ be unweighted. Then the following equivalences hold.*

1. $\psi_1^- \equiv (\psi_1^-)^+$,
2. $\psi_1^+ \wedge \psi_2^+ \equiv (\psi_1 \wedge \psi_2)^+$.

Second-Order Universal Quantification Let $\psi \in \mathsf{MSO}(\mathcal{K}, T\Sigma^*, \mathcal{F})$ be syntactically unambiguous of bounded variability b and assume $\varphi = \forall X.\psi$. Hence, ψ is of the form $\zeta \wedge (\varphi_b)^+$ for some syntactically unambiguous ζ. Note that X does not occur in φ_b. Hence, we have $\forall X.(\zeta \wedge (\varphi_b)^+) \equiv \forall X.\zeta \wedge (\varphi_b)^+$, and thus φ is also syntactically unambiguous of bounded variability b. We consider the case where ζ is of the form η^+ for some unweighted $\eta \in \mathsf{MSO}(\mathcal{K}, T\Sigma^*, \mathcal{F})$. The other cases can be reduced to this case. By definition of syntactically unambiguity and Lemma 5.24, we obtain

$$\forall X.\eta^+ \wedge (\varphi_b)^+ \equiv (\forall X.\eta)^+ \wedge (\varphi_b)^+ \equiv (\forall X.\eta \wedge \varphi_b)^+.$$

By Lemma 5.23, $[\![(\forall X.\eta \wedge \varphi_b)^+]\!] = 1_{L(\forall X.\eta \wedge \varphi_b)}$. We also have

$$L(\forall X.\eta \wedge \varphi_b) = L(\forall X.\eta) \cap L(\varphi_b) = L(\forall X.\eta) \cap T_b\Sigma^*.$$

By Theorem 5.2, $L(\forall X.\zeta)$ is TA-recognizable over $\Sigma_{\mathsf{Free}(\varphi)}$. Hence, by the first claim of Prop. 5.21, there is a deterministic timed automaton $\mathcal{A}$ such that $L(\mathcal{A}) = L(\forall X.\zeta \wedge \varphi_b)$. Let $\mathcal{A}_\varphi$ be the weighted timed automaton obtained from $\mathcal{A}$ as defined in the proof of Lemma 3.9.1. Then $\mathcal{A}_\varphi$ satisfies conditions 1. to 3. of Theorem 5.22.

For proving Theorem 5.22 for first-order universal quantification, we have to consider a modification of Lemma 5.15.

Lemma 5.25. *For every unambiguously TA-recognizable timed language $L \subseteq T\Sigma^*$ over Σ, there is an unambiguous timed automaton $\mathcal{A}'$ over Σ such that $L(\mathcal{A}') = L$ and for each location l in $\mathcal{A}'$ there is a unique $a \in \Sigma$ such that every edge $(l, a', \phi, \lambda, l')$ in $\mathcal{A}'$ satisfies $a' = a$.*

PROOF (SKETCH). The construction of $\mathcal{A}'$ is very similar to that in the proof of Lemma 5.15. However, for maintaining unambiguity of $\mathcal{A}'$, we let $\mathcal{L}'_f$ be a singleton set containing a *new* location l_f, and we add new edges of the form $(l, a, \phi, \lambda, l_f)$ for each $(l, a, \phi, \lambda, l')$ such that $l' \in \mathcal{L}_f$. This must be done to guarantee the uniqueness of the successful runs, because if we let $\mathcal{L}'_f = \mathcal{L}_f \times \Sigma$ (as in the proof of Lemma 5.15), we could not conclude that the last location (l, a) of a successful run is uniquely determined by the subsequent letter as it is for the other locations in the run. ■

First-Order Universal Quantification The proof is along the lines of the proof of Theorem 5.9. However, we have to show that $\widetilde{L}$ is *unambiguously* TA-recognizable in order to apply Lemma 5.25.

Let $\psi \in \mathsf{MSO}(\mathcal{K}, T\Sigma^*, \mathcal{F})$ be almost unambiguous over y of bounded variability b and assume $\varphi = \forall y.\psi$. We may assume that ψ is of the form

$$\psi = \bigvee_{1 \le j \le n} f_j(y) \wedge k_j \wedge \psi_j^+ \wedge (\varphi_b)^+$$

where $n \in \mathbb{N}$, $k_j \in K$, $f_j \in \mathcal{F}$, unweighted $\psi_j \in \mathsf{MSO}(\mathcal{K}, T\Sigma^*, \mathcal{F})$ for each $j \in \{1, ..., n\}$.

Let $\mathcal{W} = \mathsf{Free}(\psi)$ and $\mathcal{V} = \mathsf{Free}(\varphi) = \mathcal{W} \backslash \{y\}$. For each $i \in \{1, ..., n\}$, we have $\psi_j^+ \wedge (\varphi_b)^+ \equiv (\psi_j \wedge \varphi_b)^+$ by Lemma 5.24, and thus $[\![\psi_j^+ \wedge (\varphi_b)^+]\!] = 1_{L(\psi_j \wedge \varphi_b)}$ by Lemma 5.23. We define $\widetilde{L} \subseteq T((\Sigma^{(n)})_\mathcal{V})^*$ to be the set of timed words $((\bar{a}, \mu, \sigma), \bar{t})$ in $T((\Sigma^{(n)})_\mathcal{V})^*$ such that $((\bar{a}, \sigma), \bar{t}) \in N_\mathcal{V}$, and for all $i \in \mathsf{dom}(\bar{a}, \bar{t})$ and $j \in \{1, ..., n\}$ we have

$$\mu(i) = j \text{ implies } ((\bar{a}, \sigma[y \to i]), \bar{t}) \models \psi_j \wedge \varphi_b$$

We prove that $\widetilde{L}$ is unambiguously TA-recognizable. For this, consider the formula $\zeta \in \mathsf{MSO}(T(\Sigma^{(n)})^*)$

$$\zeta = \forall y. \bigwedge_{1 \le j \le n} \bigwedge_{a \in \Sigma} \left(P_{(a,j)}(y) \longrightarrow \widetilde{\psi_j} \wedge \widetilde{\varphi_b} \right).$$

Let $((\bar{a}, \mu, \sigma), \bar{t}) \in T((\Sigma^{(n)})_\mathcal{V})^*$ such that $((\bar{a}, \sigma), \bar{t}) \in N_\mathcal{V}$. Then we have

$$\begin{aligned}
& ((\bar{a}, \mu, \sigma), \bar{t}) \models \zeta \\
\Leftrightarrow\ & \forall i \in \mathsf{dom}(\bar{a}, \bar{t}), \forall j \in \{1, ..., n\}.\mu(i) = j \Rightarrow ((\bar{a}, \mu, \sigma[y \to i]), \bar{t}) \models \widetilde{\psi_j} \wedge \widetilde{\varphi_b} \\
\Leftrightarrow\ & \forall i \in \mathsf{dom}(\bar{a}, \bar{t}), \forall j \in \{1, ..., n\}.\mu(i) = j \Rightarrow ((\bar{a}, \sigma[y \to i]), \bar{t}) \models \psi_j \wedge \varphi_b \\
\Leftrightarrow\ & ((\bar{a}, \mu, \sigma), \bar{t}) \in \widetilde{L}.
\end{aligned}$$

Now, observe that for each $i \in \mathsf{dom}(\bar{a}, \bar{t})$ there exists some $j \in \{1, .., n\}$ and some $a \in \Sigma$ such that $P_{(a,j)}(y)$ holds. This implies that φ_b always holds. Hence, ζ is equivalent to $\zeta' \wedge \varphi_b$, where $\zeta' = \forall y. \left(\bigwedge_{1 \le j \le n} \bigwedge_{a \in \Sigma} (P_{(a,j)}(y) \longrightarrow \widetilde{\psi_j}) \right)$. Now, Theorem 5.2 implies that $L(\zeta')$ is TA-recognizable over $(\Sigma^{(n)})_\mathcal{V}$. But then, by Prop. 5.21, we know that there is a deterministic timed automaton $\widetilde{\mathcal{A}}$ recognizing $L(\zeta) = \widetilde{L}$.

From $\widetilde{\mathcal{A}}$, we construct an unambiguous timed automaton $\mathcal{A}'$ using Lemma 5.25. From this, we can define an unambiguous weighted timed automaton $\mathcal{A}$ as described in Section 5.3. Since $\mathcal{A}$ is unambiguous, we have

$$(\|\mathcal{A}\|, ((\bar{a}, \mu, \sigma), \bar{t})) = \prod_{i \in \mathsf{dom}(\bar{a}, \bar{t})} f_{\mu(i)}(t_i - t_{i-1}) \cdot k_{\mu(i)}$$

for each $((\bar{a}, \mu, \sigma), \bar{t}) \in T((\Sigma^{(n)})_\mathcal{V})^*$. Then we can proceed exactly as in Section 5.3. This finishes the proof of Theorem 5.22.

■

We define the **syntactically restricted weighted relative distance logic of bounded variability** b, denoted by $\mathsf{sR}\overleftarrow{\mathcal{L}\mathsf{d}}^{\mathsf{b}}(\mathcal{K}, \Sigma, \mathcal{F})$ to be the smallest class of formulas containing all formulas generated by the next two rules.

1. If $\varphi \in \mathsf{sRMSO}^{\mathsf{b}}(\mathcal{K}, T\Sigma^*, \mathcal{F})$, then $\varphi \in \mathsf{sR}\overleftarrow{\mathcal{L}\mathsf{d}}^{\mathsf{b}}(\mathcal{K}, \Sigma, \mathcal{F})$.
2. If $\varphi \in \mathsf{sR}\overleftarrow{\mathcal{L}\mathsf{d}}^{\mathsf{b}}(\mathcal{K}, \Sigma, \mathcal{F})$, then $\exists D.\varphi \in \mathsf{sR}\overleftarrow{\mathcal{L}\mathsf{d}}^{\mathsf{b}}(\mathcal{K}, \Sigma, \mathcal{F})$.

For the other direction, i.e., that every $\mathcal{F}$-recognizable timed series can be defined by a sentence in $\mathsf{sR}\overleftarrow{\mathcal{L}\mathsf{d}}^{\mathsf{b}}(\mathcal{K}, \Sigma, \mathcal{F})$, we can adopt the proof of Theorem 5.18 by

- considering the syntactically unambiguous version of $\psi(\bar{\mathcal{D}}, \bar{Y})$, and
- combining formulas with φ_b whenever it is needed.

We obtain a Büchi theorem for the class of $\mathcal{F}$-recognizable timed series over commutative semirings.

Theorem 5.26. *Let $\mathcal{K}$ be commutative and $\mathcal{F}$ contain $\mathbb{1}$. Each timed series $\mathcal{T} : T\Sigma^* \to K$ is $\mathcal{F}$-recognizable if and only if $\mathcal{T}$ is $\mathsf{sR}\overleftarrow{\mathcal{L}\mathsf{d}}^{\mathsf{b}}(\mathcal{K}, \Sigma, \mathcal{F})$-definable.*

Remark 5.27. The problems we solved here are due the fact that - unlike recognizable languages in the classical framework of formal languages - TA-recognizable timed languages are not determinizable (see Ex. 2.3). In the literature, one can find weighted MSO logics for other types of languages that in general are also not determinizable. In the setting of picture languages, Fichtner [85, 63, 62] introduces *first-order step functions* (rather than *recognizable step functions* as in [48]), and exploits the fact that every first-order definable picture language can be recognized by an *unambiguous* (rather than *deterministic*) picture automaton [85]. The same approach is followed by Bollig and Meinecke[20] for Mazurkiewicz traces running over directed acyclic graphs. Here, we restrict the application of first-order quantification to the subclass of timed languages with bounded variability for which it is known that it is determinizable [107]. However, there are also other characterizations of determinizable subclasses of TA-recognizable timed languages, e.g. timed languages recognizable by event-clock automata [7], or timed languages characterizable by a right morphism from the timed monoid into a bounded subset of itself [82]. Hence, it may be possible to find other restrictions on the application of the universal first-order quantifier than the restriction we propose here. Moreover, it would be interesting, also for the present context, to have an alternative characterization of unambiguously TA-recognizable timed languages.

Next, we explain how we can even skip the restriction on $\mathcal{K}$ being commutative.

In the following, let $\mathcal{K}$ be a semiring, not necessarily being commutative. We follow the approach of Droste and Gastin [49], and only present the main ideas. Commutativity of $\mathcal{K}$ mainly is needed for showing closure of the class of $\mathcal{F}$-recognizable timed series under the Hadamard product (Lemma 3.5). In the proof, we exploit the fact that the weights occuring in the runs of $\mathcal{A}_1$ *commute element-wise* with the weights occuring in the runs of $\mathcal{A}_2$. However, commutativity of $\mathcal{K}$ is a sufficient but not a necessary condition for the element-wise commutativity of weights occuring in the runs of weighted timed automata. For instance, the weights occuring in a weighted timed automaton over $\mathcal{K}$ commute element-wise with the weights occuring in a weighted timed automaton over the semiring which is generated by $\{0,1\} \subseteq K$.

For the proof of the following lemma we may proceed as in the proof of Lemma 3.5.

Lemma 5.28. *Let $\mathcal{K}_1, \mathcal{K}_2$ be two subsemirings of $\mathcal{K}$ such that $\mathcal{K}_1$ commutes element-wise with $\mathcal{K}_2$. If $\mathcal{T}_1$ is recognizable by a weighted timed automaton $\mathcal{A}_1$ over $\mathcal{K}_1$, Σ and $\mathcal{F}$, and $\mathcal{T}_2$ is recognizable by a weighted timed automaton $\mathcal{A}_2$ over $\mathcal{K}_2$, Σ and $\mathcal{F}$, and $\mathcal{A}_1$ and $\mathcal{A}_2$ are non-interfering, then $\mathcal{T}_1 \odot \mathcal{T}_2$ is $\mathcal{F}$-recognizable.*

Let $\varphi \in \overleftarrow{\mathcal{L}\mathsf{d}}(\mathcal{K}, \Sigma, \mathcal{F})$. We define $\mathsf{wgt}(\varphi) = \mathsf{wgt}_E(\varphi) \cup \mathsf{wgt}_{\mathcal{F}}(\varphi)$, where $\mathsf{wgt}_E(\varphi) = \{k : k \text{ is a subformula of } \varphi\}$ and $\mathsf{wgt}_{\mathcal{F}}(\varphi) = \{f(\delta) : f(y) \text{ is a subformula of } \varphi, \delta \in \mathbb{R}_{\geq 0}\}$. We define the **syntactically restricted auxiliary logic of bounded variability b for non-commutative semirings** $\mathsf{sRMSO}^{\mathsf{bnc}}(\mathcal{K}, T\Sigma^*, \mathcal{F})$ to be the smallest class of formulas generated by the following rules.

1. If $\varphi \in \mathsf{MSO}(\mathcal{K}, T\Sigma^*, \mathcal{F})$ is an atomic or a weighted atomic formula, then $\varphi \in \mathsf{sRMSO}^{\mathsf{bnc}}(\mathcal{K}, T\Sigma^*, \mathcal{F})$.
2. If $\varphi, \psi \in \mathsf{sRMSO}^{\mathsf{bnc}}(\mathcal{K}, T\Sigma^*, \mathcal{F})$, then $\varphi \vee \psi, \exists y.\varphi, \exists X.\varphi \in \mathsf{sRMSO}^{\mathsf{bnc}}(\mathcal{K}, T\Sigma^*, \mathcal{F})$.
3. If $\varphi \in \mathsf{sRMSO}^{\mathsf{bnc}}(\mathcal{K}, T\Sigma^*, \mathcal{F})$ is syntactically unambiguous of bounded variability b, then $\forall X.\varphi \in \mathsf{sRMSO}^{\mathsf{bnc}}(\mathcal{K}, T\Sigma^*, \mathcal{F})$.
4. If $\varphi \in \mathsf{sRMSO}^{\mathsf{bnc}}(\mathcal{K}, T\Sigma^*, \mathcal{F})$ is almost unambiguous over y of bounded variability b, then $\forall y.\varphi \in \mathsf{sRMSO}^{\mathsf{b}}(\mathcal{K}, T\Sigma^*, \mathcal{F})$.

5. If $\varphi, \psi \in \mathsf{sRMSO}^{\mathsf{bnc}}(\mathcal{K}, T\Sigma^*, \mathcal{F})$ and at least one of the following three conditions hold

   - $\mathcal{V}_f(\varphi) = \emptyset$,
   - $\mathcal{V}_f(\psi) = \emptyset$,
   - $\mathcal{V}_f(\varphi) \cap \mathcal{V}_f(\psi) = \emptyset$, $\mathcal{V}_f(\varphi) \subseteq \mathsf{Free}(\varphi)$, and $\mathcal{V}_f(\psi) \subseteq \mathsf{Free}(\psi)$,

   and

   - φ and ψ are not in the scope of a universal first-order quantifier, and
   - $\mathsf{wgt}(\varphi)$ and $\mathsf{wgt}(\psi)$ commute element-wise,

   then $\varphi \wedge \psi \wedge \bigwedge_{\substack{y \in \mathcal{V}_f(\varphi), z \in \mathcal{V}_f(\psi) \\ y \neq z}} \neg(y = z) \in \mathsf{sRMSO}^{\mathsf{bnc}}(\mathcal{K}, T\Sigma^*, \mathcal{F})$.

We use $\mathsf{sR}\overleftarrow{\mathcal{L}\mathsf{d}}^{\mathsf{bnc}}(\mathcal{K}, \Sigma, \mathcal{F})$ to denote the smallest class of formulas containing formulas generated by the next two rules.

1. If $\varphi \in \mathsf{sRMSO}^{\mathsf{bnc}}(\mathcal{K}, T\Sigma^*, \mathcal{F})$, then $\varphi \in \mathsf{sR}\overleftarrow{\mathcal{L}\mathsf{d}}^{\mathsf{bnc}}(\mathcal{K}, \Sigma, \mathcal{F})$.

2. If $\varphi \in \mathsf{sR}\overleftarrow{\mathcal{L}\mathsf{d}}^{\mathsf{bnc}}(\mathcal{K}, \Sigma, \mathcal{F})$, then $\exists D.\varphi \in \mathsf{sR}\overleftarrow{\mathcal{L}\mathsf{d}}^{\mathsf{bnc}}(\mathcal{K}, \Sigma, \mathcal{F})$.

Note that, as opposed to the other conditions, it depends on $\mathcal{K}$ and $\mathcal{F}$ whether one can check syntactically whether two given formulas φ and ψ satisfy that $\mathsf{wgt}(\varphi)$ and $\mathsf{wgt}(\psi)$ commute element-wise. For instance, if $\mathcal{K}$ is commutative, we do not need to check syntactically for element-wise commuting of the weights appearing in the formula. If on the other hand $\mathcal{K}$ is not commutative but $\mathsf{wgt}(\varphi)$ is a finite set, which is e.g. the case if there are no location weight functions occuring in φ, or $\mathcal{F}$ is the family of step functions, then we can easily check syntactically whether the weights in $\mathsf{wgt}(\varphi)$ commute element-wise. For other cases, this might not be so easy or even impossible.

The proof for the direction from definability to recognizability is analogous to the proofs for commutative semirings, using Lemma 5.28. For a proof for the direction from recognizability to definability, we must be careful with the construction of the formula $\xi(\bar{D}, \bar{Y})$ (see Theorem 5.18). Note that the constants $\mathsf{in}(l_i)$ and $\mathsf{out}(l_i)$ need not necessarily commute element-wise, and thus, $\xi(\bar{D}, \bar{Y})$ is not guaranteed to satisfy the fourth condition. In the untimed case, this can be solved using a result of Eilenberg [58], which allows us to choose an equivalent initial- and final-state-normalized weighted finite automaton in which $\mathsf{in}(q), \mathsf{out}(q) \in \{0, 1\}$ for each $q \in Q$ [49]. However, as explained in Remark 4.12, for weighted timed automata $\mathcal{A}$ over arbitrary $\mathcal{K}$ and $\mathcal{F}$, we cannot provide an initial-location-normalized weighted timed automaton $\mathcal{A}'$ with the same behaviour like $\mathcal{A}$ such that $\mathsf{in}(l) \in \{0, 1\}$ for each $l \in \mathcal{L}'$. For this reason, if $\mathcal{K}$ is not commutative, we cannot give a Büchi theorem for weighted timed automata $\mathcal{A}$ over arbitrary $\mathcal{F}$. We may consider weighted timed automata with restricted functions for entering locations of the

Weighted Timed MSO Logic	Auxiliary Logic	Syntactical Restrictions	$\mathcal{F}$-Rec.Timed Series over $\mathcal{K}$ and Σ
$\overleftarrow{\mathcal{L}\mathsf{d}}\,(\mathcal{K},\Sigma,\mathcal{F})$	$\mathsf{MSO}(\mathcal{K},T\Sigma^*,\mathcal{F})$	no restriction	none
$\mathsf{sR}\overleftarrow{\mathcal{L}\mathsf{d}}\,(\mathcal{K},\Sigma,\mathcal{F})$	$\mathsf{sRMSO}(\mathcal{K},T\Sigma^*,\mathcal{F})$	(1),(2),(3)	$\mathcal{K}$ is idempotent and commutative
$\mathsf{sR}\overleftarrow{\mathcal{L}\mathsf{d}}^{\,\mathsf{b}}(\mathcal{K},\Sigma,\mathcal{F})$	$\mathsf{sRMSO}^{\mathsf{b}}(\mathcal{K},T\Sigma^*,\mathcal{F})$	(1),(2b),(3b)	$\mathcal{K}$ is commutative
$\mathsf{sR}\overleftarrow{\mathcal{L}\mathsf{d}}^{\,\mathsf{bnc}}(\mathcal{K},\Sigma,\mathcal{F})$	$\mathsf{sRMSO}^{\mathsf{bnc}}(\mathcal{K},T\Sigma^*,\mathcal{F})$	(1),(2b),(3b),(4)	ε is excluded

(1) Restriction on conjunction.
(2) If φ contains $\forall y.\psi$, then ψ is almost unambiguous over y.
(2b) If φ contains $\forall y.\psi$, then ψ is almost unambiguous over y of bounded variability b.
(3) If φ contains $\forall X.\psi$, then ψ is unweighted.
(3b) If φ contains $\forall X.\psi$, then ψ is syntactically unambiguous over y of bounded variability.
(4) If φ contains $\varphi_1 \wedge \varphi_2$ and this is not in the scope of a universal first-order quantifier, then $\mathsf{wgt}(\varphi_1)$ and $\mathsf{wgt}(\varphi_2)$ commute element-wise.

Table 5.1: Overview of Weighted Timed MSO Logics

form $\mathsf{in} : \mathcal{L} \to \{0,1\}$ (or, equivalently, with a designated set of initial states, as in [53]). However, as mentioned in Remark 5.14, in this case a Büchi theorem can only be given for $\mathcal{F}$-recognizable timed series $\mathcal{T} : T\Sigma^+ \to K$, i.e., excluding the empty timed word.

Theorem 5.29. *Let $\mathcal{F}$ contain $\mathbb{1}$. Then a timed series $\mathcal{T} : T\Sigma^+ \to K$ is $\mathcal{F}$-recognizable if and only if $\mathcal{T}$ is $\mathsf{sR}\overleftarrow{\mathcal{L}\mathsf{d}}^{\,\mathsf{bnc}}(\mathcal{K},\Sigma,\mathcal{F})$-definable.*

We note that we may also come up with a Büchi theorem for $\mathcal{F}$-recognizable timed series if $\mathcal{F}$ is the family of step functions, because for this family of functions we can present an initial-location-normalization such that $\mathsf{in}(l) \in \{0,1\}$ (see Remark 4.12). In Table 5.1, we give a summary of the different weighted timed MSO logics we defined in this article. In the third column, we list the syntactical restrictions of the logic. In the fourth column, we specify the class of $\mathcal{F}$-recognizable timed series such that a Büchi theorem holds. Note that we always assume $\mathbb{1} \in \mathcal{F}$.

5.6 Conclusion

We have presented the weighted relative distance logic, which is - at least to our knowledge - the first MSO logic allowing for the description of both timed and quantitative properties. On the one hand, our logic may be used as a new tool for specifying properties. It sometimes may be easier to specify properties in terms of logic rather than by automata devices. As an example, consider the simple formula given in Ex. 5.5 and the weighted timed automaton of Ex. 3.2, both of which represent the same behaviour. On the other hand, this logic gives rise to some interesting new directions in future research work. For instance, Wilke [107] showed that some real-time temporal logics are effectively embeddable into the relative distance logic. All his constructions for obtaining a Büchi theorem are effective. By the decidability of the emptiness problem for timed automata (cf. Theorem 2.5), one can thus conclude that these real-time temporal logics have a decidable theory. This gives rise to the question whether one can obtain similar results for weighted extensions of real-time temporal logics.

We also would like to mention that our logic and constructions follow the ideas of the work of Droste and Gastin [49] and thus we keep the spirit of the untimed theory. However, we additionally allowed functions from the family as atomic formulas, which complicates most of the proofs, first and foremost the proof for showing closure of the class of recognizable timed series under first-order universal quantification. Moreover, we had to deal with the problem that - unlike finite automata - timed automata are not determinizable in general. Similarly to the Kleene-Schützenberger theorem, one may also ask for a Büchi theorem for weighted timed automata on infinite words. For this, one may consider the work by Droste and Rahonis [54] on weighted logics on infinite words. Last but not least, it is an interesting question to which class of automata the unrestricted logic $\overleftarrow{\mathcal{L}\mathsf{d}}\,(\mathcal{K}, \Sigma, \mathcal{F})$ is expressively equivalent.

6 Supports and Timed Cut Languages

6.1 Introduction

In this chapter, we aim to shed light on the *supports* of timed series, consisting of all timed words which are not mapped to zero. Within the theory of weighted finite automata, supports have been extensively studied (see e.g. [19, 98]). For instance, for large classes of semirings, the support of a WFA-recognizable series is known to be recognizable by a finite automaton (see e.g. Theorem 2.8). This implies the decidability of weighted versions of some of the fundamental decision problems of formal language theory as e.g. the emptiness problem. For the semiring of the reals with addition and multiplication and the family of linear functions, we can show that the question, whether the support of a recognizable timed series is empty, is decidable. This and results obtained in previous chapters imply that it is decidable whether two recognizable timed series are equal. For this, we use that two timed series $\mathcal{T}_1$ and $\mathcal{T}_2$ are equal if and only if the difference $\mathcal{T}_1 - \mathcal{T}_2$ equals the constant zero timed series (or, equivalently, the support of this difference is empty). For recognizable timed series $\mathcal{T}_1$ and $\mathcal{T}_2$ over the semiring of the reals with addition and multiplication we can construct a weighted timed automaton recognizing the difference $\mathcal{T}_1 - \mathcal{T}_2$, because this semiring is a field. Note that for recognizable timed series over e.g. the Boolean semiring this construction is not possible. Indeed, the corresponding problem of equality of two TA-recognizable timed languages is undecidable.

We also want to investigate the TA-recognizability of *timed cut languages*. These are sets of timed words which are assigned a weight smaller than (or greater than, respectively) a given value. Both supports and timed cut languages may be useful in analysing real-time systems. For instance, we may be interested in whether the set of timed words whose weight under a weighted timed automaton is not exceeding a given value satisfies a specification. Problems like this can be solved by using the automata-theoretic methods presented in this chapter. Moreover, from the decidability results we are going to present here, we immediately obtain decidability results for the weighted relative distance logic introduced in Sect. 5.

Here, we focus on $\mathcal{F}$-recognizable timed series over semirings having weights in the reals, where $\mathcal{F}$ is the family of linear functions. This setting allows for interesting applications in the theory of weighted timed automata [17, 8, 22]. Many algorithms for decision problems in the theory of weighted finite automata rely on the fact that the set of weights occuring in a weighted finite automaton is finite. However, in a weighted timed automaton with linear weight functions this is not the case. Thus, the main challenge is to deal with the potentially infinite number of weights occuring in a weighted timed automaton. We can show that for some problems and semirings it is not necessary to consider the *exact* weights of the transitions participating in a run. Besides linear functions, we consider weighted timed automata over step functions for which the problem of infinite weights does not occur. For this kind of functions we can show that most of the results of the untimed setting can be carried over to the timed setting.

6.2 Recognizability of Supports of Recognizable Timed Series

In this section, we fix a semiring $\mathcal{K}$, an alphabet Σ and a family $\mathcal{F}$ of functions from $\mathbb{R}_{\geq 0}$ to K. We study whether the support of an $\mathcal{F}$-recognizable timed series is a TA-recognizable timed language. Recall that in the untimed setting the support of every WFA-recognizable series over a *positive* semiring is recognizable by a finite automaton (Theorem 2.8). The main idea for the construction of the finite automaton is to remove edges with weight 0 from the given weighted finite automaton and ignore the remaining weights. For the timed setting, this idea can more or less be adopted by removing not only the edges with weight 0 but also edges whose source location is assigned a weight function that maps *all* time delays to 0. However, for certain positive semirings and families the method only works for *strictly* monotonic $\mathcal{F}$-recognizable timed series, as the following example illustrates.

Example 6.1. Let $\mathcal{K}$ be the semiring of the real numbers with ordinary addition and multiplication, $\mathcal{F}$ be the family of linear functions and $\mathcal{A}$ be a weighted timed automaton over $\mathcal{K}$, Σ and $\mathcal{F}$. Let $w \in T\Sigma^*$. Then, whenever there is some $i \in \mathsf{dom}(w)$ with $t_i - t_{i-1} = 0$, we have $(\|\mathcal{A}\|, w) = 0$. So, the zero behaviour of $\mathcal{A}$ with respect to w is due to the form of w and has nothing to do with the particular definition of $\mathcal{A}$.

For this reason, in the next lemma we distinguish between two cases.

Theorem 6.2. *Let $\mathcal{K}$ be positive.*

1. *Assume that for all $f \in \mathcal{F}$ we have either $f(\delta) \neq 0$ for all $\delta \in \mathbb{R}_{\geq 0}$ or $f(\delta) = 0$ for all $\delta \in \mathbb{R}_{\geq 0}$. Then the support of each $\mathcal{F}$-recognizable timed series over $\mathcal{K}$ and Σ is TA-recognizable over Σ.*

2. *Assume that for all $f \in \mathcal{F}$ we have either $f(\delta) \neq 0$ for all $\delta \in \mathbb{R}_{\geq 0}\backslash\{0\}$ or $f(\delta) = 0$ for all $\delta \in \mathbb{R}_{\geq 0}\backslash\{0\}$. Then the support of each strictly monotonic and $\mathcal{F}$-recognizable timed series over $\mathcal{K}$ and Σ is strictly monotonic and TA-recognizable over Σ.*

PROOF. 1. Let $\mathcal{A} = (\mathcal{L}, \mathcal{C}, E, \mathsf{in}, \mathsf{out}, \mathsf{ewt}, \mathsf{lwt})$ be a weighted timed automaton over $\mathcal{K}$, Σ and $\mathcal{F}$, where $\mathcal{F}$ satisfies the assumption stated. We define

$E' = \{e \in E : \mathsf{ewt}(e) \neq 0, \mathsf{lwt}(\mathsf{source}(e))(\delta) \neq 0 \text{ for each } \delta \in \mathbb{R}_{\geq 0}\}$. We further define $\mathcal{L}_0 = \{l \in \mathcal{L} : \mathsf{in}(l) \neq 0\}$ and $\mathcal{L}_f = \{l \in \mathcal{L} : \mathsf{out}(l) \neq 0\}$. We show that for the timed automaton $\mathcal{A}' = (\mathcal{L}, \mathcal{L}_0, \mathcal{L}_f, \mathcal{C}, E')$ we have $L(\mathcal{A}') = \mathsf{supp}(\|\mathcal{A}\|)$.

First, let $w \in \mathsf{supp}(\|\mathcal{A}\|)$, i.e., $(\|\mathcal{A}\|, w) \neq 0$. Since $\mathcal{K}$ is zero-sum free, there must be a run r of $\mathcal{A}$ on w such that $\mathsf{rwt}(r) \neq 0$. Let r be of the form $(l_0, \nu_0) \xrightarrow{\delta_1} \xrightarrow{e_1} \ldots \xrightarrow{\delta_{|w|}} \xrightarrow{e_{|w|}} (l_{|w|}, \nu_{|w|})$. Then, we have $\mathsf{in}(l_0) \cdot (\prod_{1 \leq i \leq |w|} \mathsf{lwt}(l_{i-1})(\delta_i) \cdot \mathsf{ewt}(e_i)) \cdot \mathsf{out}(l_{|w|}) \neq 0$. Since 0 is absorbing, each of the factors must be different from 0. In particular, we must have $\mathsf{in}(l_0) \neq 0$ and thus can conclude $l_0 \in \mathcal{L}_0$. Similarly, $\mathsf{out}(l_{|w|}) \neq 0$ and thus $l_{|w|} \in \mathcal{L}_f$. Further, we have $\mathsf{lwt}(l_{i-1})(\delta_i) \neq 0$ for each $i \in \mathsf{dom}(w)$ and thus, by the restriction we put on $\mathcal{F}$, we know that $\mathsf{lwt}(l_{i-1})(\delta) \neq 0$ for each $\delta \in \mathbb{R}_{\geq 0}$. This, together with $\mathsf{ewt}(e_i) \neq 0$ implies $e_i \in E'$ for every $i \in \mathsf{dom}(w)$ and r is a successful run of $\mathcal{A}'$ on w, which implies $w \in L(\mathcal{A}')$.

Second, let $w \in L(\mathcal{A}')$. Hence, there must be a successful run r of $\mathcal{A}'$ on w. Let r be of the form $(l_0, \nu_0) \xrightarrow{\delta_1} \xrightarrow{e_1} \ldots \xrightarrow{\delta_{|w|}} \xrightarrow{e_{|w|}} (l_{|w|}, \nu_{|w|})$. Clearly, r must also be a run of $\mathcal{A}$ on w. By definition of $\mathcal{L}_0$ and $\mathcal{L}_f$, we have $\mathsf{in}(l_0) \neq 0$ and $\mathsf{out}(l_{|w|}) \neq 0$. By definition of E', for every $i \in \mathsf{dom}(w)$, we have $\mathsf{ewt}(e_i) \neq 0$ and l_{i-1} satisfies $\mathsf{lwt}(l_{i-1})(\delta) \neq 0$ for each $\delta \in \mathbb{R}_{\geq 0}$. Hence, $\mathsf{lwt}(l_{i-1})(\delta_i) \neq 0$ for every $i \in \mathsf{dom}(w)$. Now, $\mathcal{K}$ is zero-divisor free and thus we must have $\mathsf{rwt}(r) \neq 0$. The zero-sum freeness of $\mathcal{K}$ implies $(\|\mathcal{A}\|, w) \neq 0$ and thus $w \in \mathsf{supp}(\|\mathcal{A}\|)$.

2. The proof can be done analogously. Additionally, in the proof of $L(\mathcal{A}') \subseteq \mathsf{supp}(\|\mathcal{A}\|)$ we use the fact that zero time delays are not allowed to conclude that $\mathsf{lwt}(l_{i-1})(\delta_i) \neq 0$.

■

As a consequence, the support of the behaviour of each weighted timed automaton over the widely-used [17, 8, 22] setting of the min-plus-semiring and the family of linear functions is TA-recognizable.

In the past few years, weighted timed automata with multiple prices have attracted interest [24, 81]. These may be modelled using the direct product of e.g. the min-plus-semiring with itself and componentwise defined linear functions. Unfortunately, while still being zero-sum free, the direct product of two positive semirings is not necessarily zero-divisor free and thus Theorem 6.2 cannot be applied. Very recently, Kirsten [75] came up with a result supplementing Theorem 2.8, see Theorem 2.9 given in Sect. 2.3. Kirsten's proof method relies on the fact that the set of disjoint weights occurring in any of the runs of the weighted finite automaton is finite. However, for weighted timed automata over e.g. the family of linear functions this is not the case. Nevertheless, for certain semirings we can adapt the proof by exploiting the fact that the *exact* weights emerging from staying in a location are not crucial for deciding whether the running weight of a run equals 0 or not, but rather it is sufficient to consider the semiring coefficients that the time delay is multiplied with, as the following example shows.

Example 6.3. Let $\mathcal{A} = (\mathcal{L}, \mathcal{C}, E, \mathsf{in}, \mathsf{out}, \mathsf{ewt}, \mathsf{lwt})$ be a weighted timed automaton over the min-plus-semiring and the family of linear functions. For the sake of simplicity, in this example we assume that $\mathsf{in}(l) = \mathsf{out}(l) = 0$ for each $l \in \mathcal{L}$ and $\mathsf{ewt}(e) = 0$ for each $e \in E$. Let $w \in T\Sigma^*$. Then for each run r of $\mathcal{A}$ on w we have $\mathsf{rwt}(r) = \infty$ if and only if there is some $i \in \mathsf{dom}(w)$ such that $\mathsf{lwt}(l_{i-1})(\delta_i) = \infty$. However, for every location l and time delay δ, we have $\mathsf{lwt}(l)(\delta) = \infty$ if and only if $\mathsf{lwt}(l)(\delta') = \infty \cdot \delta'$ for each $\delta' \in \mathbb{R}_{\geq 0}$. Hence, the exact time delays are not important but it is sufficient to consider the coefficients of the weight functions of the participating locations in a run to decide whether $\mathsf{rwt}(r) = \infty$. For weighted timed automata over the direct product of the min-plus-semiring with itself and the family of componentwise defined linear functions, we have a similar result, namely $\mathsf{rwt}(r) = (\infty, \infty)$ if and only if there are $i, j \in \mathsf{dom}(w)$ such that $\mathsf{lwt}(l_{i-1})(\delta_i) = (k, \infty)$ and $\mathsf{lwt}(l_{j-1})(\delta_j) = (\infty, k')$, for some $k, k' \in K$. Again, the exact time delays are not important.

This leads us to the following result.

Theorem 6.4.

1. *Let $\mathcal{K}$ be commutative and zero-sum free and let $\mathcal{F}$ be the family of step functions. Then the support of each $\mathcal{F}$-recognizable timed series over $\mathcal{K}$ and Σ is TA-recognizable over Σ.*

2. *Let $\mathcal{K}$ be one of the following semirings*

   a) the min-plus-semiring,

   b) the max-plus-semiring,

   c) the min-max-semiring,

   or the direct product of any two of these semirings, and let $\mathcal{F}$ be the family of (componentwise defined) linear functions. Then the support of each $\mathcal{F}$-recognizable timed series over $\mathcal{K}$ and Σ is TA-recognizable over Σ.

3. *Let $\mathcal{K}$ be the semiring of the positive real numbers together with ordinary addition and multiplication or the direct product of this with itself, and let $\mathcal{F}$ be the family of (componentwise defined) linear functions. Then the support of each strictly monotonic and $\mathcal{F}$-recognizable timed series over $\mathcal{K}$ and Σ is strictly monotonic and TA-recognizable over Σ.*

Before we present the proof of this theorem, we recall some notions and results introduced by Kirsten [75]. For $n \in \mathbb{N}$, given two tuples $\bar{y}, \bar{z} \in \mathbb{N}^n$, we let $\bar{y} \leq \bar{z}$ if $y_i \leq z_i$ for every $i \in \{1, ..., n\}$. Given some subset $M \subseteq \mathbb{N}^n$, we denote by $\mathrm{Min}(M)$ the set of all minimal tuples of M, formally $\mathrm{Min}(M) = \{\bar{y} \in M : \text{ for every } \bar{z} \in M, \bar{z} \leq \bar{y} \text{ implies } \bar{y} = \bar{z}\}$. Given some $\bar{y} \in \mathbb{N}^n$ and some $z \in \mathbb{N}$, we denote by $\lfloor \bar{y} \rfloor_z$ the tuple defined by $(\lfloor \bar{y} \rfloor_z)_i = \min\{y_i, z\}$ for every $i \in \{1, ..., n\}$. Let $(K, \cdot, 1)$ be a monoid with a zero element 0 such that $0 \cdot k = k \cdot 0 = 0$ for each $k \in K$.

Let $V = (c_1, ..., c_n) \in K^n$ be a tuple. We define the monoid morphism $\eta : (\mathbb{N}^n, +, 0^n) \to (K, \cdot, 1)$ by $\eta(\bar{y}) = c_1^{y_1} \cdot ... \cdot c_n^{y_n}$ for every $\bar{y} = (y_1, ..., y_n) \in \mathbb{N}^n$. By Dickson's Lemma [43], the set $\mathrm{Min}(\eta^{-1}(0))$ must be finite. By $\mathrm{dg}(V)$ we denote the least non-negative integer such that $\mathrm{Min}(\eta^{-1}(0))$ is a subset of $\{0, ..., \mathrm{dg}(V)\}^n$.

We define a partial mapping $\pm : \{0, ..., \mathrm{dg}(V)\}^n \times K \to \{0, ..., \mathrm{dg}(V)\}^n$. For this, let $\bar{y} \in \{0, ..., \mathrm{dg}(V)\}^n$ and $k \in K$. We define $\bar{y} \pm k$ if k occurs in V. So assume there is some unique $i \in \{1, ..., n\}$ such that $c_i = k$. Then, $\bar{z} \in \{0, ..., \mathrm{dg}(V)\}^n$ is defined by

$$z_j = \begin{cases} y_j + 1 & \text{if } j = i, \\ y_j & \text{otherwise.} \end{cases}$$

We define $\bar{y} \pm k = \lfloor \bar{z} \rfloor_{\mathrm{dg}(V)}$. Let $m, m' \in \mathbb{N}$ and $k_1, ..., k_m, k'_1, ..., k'_{m'} \in K$. The *zero generation problem* (ZGP) means to decide whether there exists some k in the monoid generated by $k'_1, ..., k'_{m'}$ such that $k_1 \cdot ... \cdot k_m \cdot k = 0$.

Lemma 6.5 ([75]). *If the ZGP is decidable in $\mathcal{K}$, then we can effectively compute $dg(V)$ from V.*

Lemma 6.6 ([75]). *For every $\bar{y} \in \mathbb{N}^n$ we have $\eta(\bar{y}) = 0$ if and only if $\eta(\lfloor \bar{y} \rfloor_{dg(V)}) = 0$.*

The following observation is crucial for applying Kirsten's proof method irrespective of the fact that $\mathsf{wgt}(\mathcal{A})$ may be infinite.

Lemma 6.7. *Let $\mathcal{K}$ be one of the following semirings*

1. *min-plus-semiring,*
2. *max-plus-semiring,*
3. *min-max-semiring,*

or the direct product of any two of these semirings. Furthermore, let $n \in \mathbb{N}$ and $k_i \in K, \delta_i \in \mathbb{R}_{\geq 0}$ for each $i \in \{1, ..., n\}$. Then we have

$$\prod_{1 \leq i \leq n} k_i \cdot \delta_i = 0 \quad \textit{if and only if} \quad \prod_{1 \leq i \leq n} k_i = 0.$$

PROOF. We let $\mathcal{K}$ be the direct product of the min-plus-semiring with itself. The proof for the other semirings can be adapted. So let $n \in \mathbb{N}$ and $k_i \in K, \delta_i \in \mathbb{R}_{\geq 0}$ for each $i \in \{1, ..., n\}$. Then we have

$$\begin{aligned} & \sum_{1 \leq i \leq n} k_i \cdot \delta_i = (\infty, \infty) \\ \Leftrightarrow\ & \text{there exist } i, j \in \{1, ..., n\} \text{ such that } k_i = (k, \infty) \text{ and } k_j = (\infty, k') \text{ for some } k, k' \in K \\ \Leftrightarrow\ & \sum_{1 \leq i \leq n} k_i = (\infty, \infty). \end{aligned}$$

■

For proving the effectiveness of the construction, we use the next Lemma.

Lemma 6.8. *Let $\mathcal{K}$ be one of the following semirings*

1. *min-plus-semiring,*
2. *max-plus-semiring,*
3. *min-max-semiring,*

or the direct product of any two of these semirings. Then the ZGP is decidable.

PROOF. We show the proof for the min-plus-semiring. So let $k_1, ..., k_m, k'_1, ..., k'_{m'} \in \mathbb{R}_{\geq 0} \cup \{\infty\}$ for some $m, m' \in \mathbb{N}$. We want to decide whether there is some k in the monoid generated by $k'_1, ..., k'_{m'}$ such that $k_1 + ... + k_m + k = \infty$. Clearly, this is the case if and only if there is some $i \in \{1, ..., m\}$ such that $k_i = \infty$ or there is some $i \in \{1, ..., m'\}$ such that $k'_i = \infty$. ∎

PROOF OF THEOREM 6.4. We present the proof for the second claim. The proof of 1. can be done analogously to the proof by Kirsten [75]. The proof of 3. can be done as the proof of 2., and by additionally exploiting the fact that zero time delays are not allowed.

So let $\mathcal{K}$ be the direct product of the min-plus-semiring with itself. The proof for the other semirings can be done analogously. Let $\mathcal{F}$ be the family of componentwise defined linear functions and $\mathcal{T} : T\Sigma^* \to K$ be an $\mathcal{F}$-recognizable timed series. Then there is a weighted timed automaton $\mathcal{A} = (\mathcal{L}, \mathcal{C}, E, \mathsf{in}, \mathsf{out}, \mathsf{ewt}, \mathsf{lwt})$ over $\mathcal{K}$, Σ and $\mathcal{F}$ such that $\|\mathcal{A}\| = \mathcal{T}$. For each $l \in \mathcal{L}$, we let $k_l \in (\mathbb{R}_{\geq 0} \cup \{\infty\}) \times (\mathbb{R}_{\geq 0} \cup \{\infty\})$ denote the coefficient used in the weight function of l, i.e., if $\mathsf{lwt}(l)(\delta) = (k \cdot \delta, k' \cdot \delta)$ for each $\delta \in \mathbb{R}_{\geq 0}$ and some $k, k' \in \mathbb{R}_{\geq 0} \cup \{\infty\}$, then $k_l = (k, k')$. Let $n \in \mathbb{N}$ and $V = (c_1, ..., c_n) \in ((\mathbb{R}_{\geq 0} \cup \{\infty\}) \times (\mathbb{R}_{\geq 0} \cup \{\infty\}))^n$ such that

- for every $e \in E$ there is exactly one $i \in \{1, ..., n\}$ satisfying $c_i = \mathsf{ewt}(e)$,
- for every $l \in \mathcal{L}$ there is exactly one $i \in \{1, ...n\}$ satisfying $c_i = \mathsf{in}(l)$, there is exactly one $i \in \{1, ...n\}$ satisfying $c_i = \mathsf{out}(l)$, and there is exactly one $i \in \{1, ..., n\}$ satisfying $c_i = k_l$.

Furthermore, for every $i \in \{1, ..., n\}$, c_i either is the weight of an edge $e \in E$, the weight for entering or leaving some location $l \in \mathcal{L}$ or it is the coefficient k_l with which a time delay is multiplied in some location $l \in \mathcal{L}$. Notice that by Lemmas 6.8 and 6.5, $\mathrm{dg}(V)$ is effectively computable. We define the timed automaton $\mathcal{A}' = (\mathcal{L}', \mathcal{L}'_0, \mathcal{L}'_f, \mathcal{C}, E')$ as follows:

- $\mathcal{L}' = \mathcal{L} \times \{0, ..., \mathrm{dg}(V)\}^n$,
- $(l, \bar{y}) \in \mathcal{L}'_0$ if and only if there is some $i \in \{1, ..., n\}$ such that $y_i = 1$ and $\mathsf{in}(l) = c_i$ and for all $j \in \{1, ..., n\}$ with $j \neq i$ we have $y_j = 0$,

- $(l, \bar{y}) \in \mathcal{L}'_f$ if and only if $\eta(\bar{y} \pm \mathsf{out}(l)) \neq (\infty, \infty)$,
- $((l, \bar{y}), a, \phi, \lambda, (l', \bar{z})) \in E'$ if and only if there exists an edge $(l, a, \phi, \lambda, l') \in E$ such that $\bar{y} \pm k_l \pm \mathsf{ewt}(l, a, \phi, \lambda, l') = \bar{z}$. We say that $((l, \bar{y}), a, \phi, \lambda, (l', \bar{z}))$ stems from $(l, a, \phi, \lambda, l')$.

Next, we show that $L(\mathcal{A}') = \mathsf{supp}(\|\mathcal{A}\|)$. For this, let $w \in T\Sigma^*$ and assume $w \in L(\mathcal{A}')$. Then there is a successful run $((l_0, \bar{y}_0), \nu_0) \xrightarrow{\delta_1} \xrightarrow{e'_1} \dots \xrightarrow{\delta_{|w|}} \xrightarrow{e'_{|w|}} ((l_{|w|}, \bar{y}_{|w|}), \nu_{|w|})$ of $\mathcal{A}'$ on w. For every $j \in \mathsf{dom}(w)$, let $e_j \in E$ such that e'_j stems from e_j. Clearly, $r = (l_0, \nu_0) \xrightarrow{\delta_1} \xrightarrow{e_1} \dots \xrightarrow{\delta_{|w|}} \xrightarrow{e_{|w|}} (l_{|w|}, \nu_{|w|})$ is a run of $\mathcal{A}$ on w. First of all, we have $(l_0, \bar{y}_0) \in \mathcal{L}'_0$ and thus, by definition of $\mathcal{L}'_0$, $\eta(\bar{y}_0) = \mathsf{in}(l_0)$. For $j \in \{0, ..., |w|\}$, let $\bar{z}_j \in \mathbb{N}^n$ be the tuple such that for every $i \in \{1, ..., n\}$, $z_{j,i}$ is the number of occurences of c_i among $\mathsf{in}(l_0), k_{l_0}, \mathsf{ewt}(e_1), ..., k_{l_{j-1}}, \mathsf{ewt}(e_j)$. In particular, $\bar{z}_0 = \bar{y}_0$. Let $\bar{z} \in \mathbb{N}^n$ such that for every $i \in \{1, ..., n\}$, z_i is the number of occurrences of c_i among $\mathsf{in}(l_0), k_{l_0}, \mathsf{ewt}(e_1), ..., k_{l_{|w|-1}}, \mathsf{ewt}(e_{|w|}), \mathsf{out}(l_{|w|})$. Clearly, $\eta(\bar{z}) = \mathsf{rwt}(r')$. By induction, one can show that for every $j \in \{0, ..., |w| - 1\}$, $\bar{y}_j = \lfloor \bar{z}_j \rfloor_{\mathrm{dg}(V)}$ and $\bar{y}_{|w|} \pm \mathsf{out}(l_{|w|}) = \lfloor \bar{z} \rfloor_{\mathrm{dg}(V)}$. As $(l_{|w|}, \bar{y}_{|w|}) \in \mathcal{L}'_f$, we must have $\eta(\bar{y}_{|w|} \pm \mathsf{out}(l)) \neq (\infty, \infty)$. Hence, we also have $\eta(\lfloor \bar{z} \rfloor_{\mathrm{dg}(V)}) \neq (\infty, \infty)$. By Lemma 6.6, we obtain $\eta(\bar{z}) \neq (\infty, \infty)$. This together with Lemma 6.7 implies $\mathsf{rwt}(r) \neq (\infty, \infty)$. Since $\mathcal{K}$ is zero-sum free, we get $w \in \mathsf{supp}(\|\mathcal{A}\|)$.

The proof for the other direction can be done analogously: for $w \in \mathsf{supp}(\|\mathcal{A}\|)$, there must be a run of $\mathcal{A}$ on w, for which a corresponding successful run of $\mathcal{A}'$ can be constructed according to the definition of E'. For proving $w \in L(\mathcal{A}')$ one can show that this run must be successful using fairly the same methods as in the proof for the other direction. ■

We combine the results of this section with some decidability results for timed automata, namely the decidability of the emptiness problem (see Theorem 2.5) and decidability of the universality problem for single-clock timed automata (see Theorem 2.6). The latter can be done since none of the constructions in the proofs of the previous lemmas increase the number of clock variables in the timed automaton $\mathcal{A}'$. As a result, we obtain decidability results for weighted versions of the classical emptiness problem and the universality problem.

Corollary 6.9. *The emptiness of the support of an $\mathcal{F}$-recognizable timed series over a semiring and a family as specified in Theorems 6.2 and 6.4 is decidable. Moreover, it is decidable, for a given timed series that can be recognized by a single-clock weighted timed automaton over such a semiring and family, whether its support is universal.*

The weighted version of the emptiness problem, called the *empty support problem* is also subject of the next section.

6.3 The Empty Support Problem for Recognizable Timed Series over Fields

We consider $\mathcal{F}$-recognizable timed series over fields, i.e., semirings where $(K, +, 0)$ is a group and $(K\backslash\{0\}, \cdot, 1)$ is a commutative group. Recall that already in the untimed setting there are WFA-recognizable series over fields for which the support is not recognizable by a finite automaton (see Example 2.10). Yet, by Theorem 2.11 it is decidable whether the support of a WFA-recognizable series over a field is empty. In this section, we show that this also holds for $\mathcal{F}$-recognizable timed series over some fields and families $\mathcal{F}$. The main idea is to reduce the problem to the corresponding problem for the class of WFA-recognizable series. For this, we construct weighted versions of the classical region automaton [6].

Let $\mathcal{K}$ be a field and $\mathcal{F}$ be the family of step functions. Given a weighted timed automaton $\mathcal{A} = (\mathcal{L}, \mathcal{C}, E, \mathsf{in}, \mathsf{out}, \mathsf{ewt}, \mathsf{lwt})$ over $\mathcal{K}$, Σ and $\mathcal{F}$, we define the weighted finite automaton $R_{\mathsf{step}}(\mathcal{A}) = (Q, \Delta, \mathsf{in}', \mathsf{out}', \mathsf{wt})$ over $\mathcal{K}$ and $\mathcal{I} \times \Sigma$, where $(Q, \Delta, \mathsf{in}', \mathsf{out}')$ is the region automaton defined in Sect. 2.1, and the weight function wt is defined as follows: let $t = \big((l, r), (I, a), (l', r')\big) \in \Delta$ stem from e and δ for some $e \in E$ and $\delta \in I$. Further assume that $\mathsf{lwt}(l)$ is of the form $\sum_{1 \leq i \leq n} \alpha_i \cdot \chi_{A_i}(\delta')$ for every $\delta' \in \mathbb{R}_{\geq 0}$. By the definition of step functions, there must be some $i \in \{1, ..., n\}$ such that $\delta \in A_i$, and we have $\delta \notin A_j$ for each $j \in \{1, ..., n\}$ with $j \neq i$. Note that i is uniquely determined. Then, we define $\mathsf{wt}(t) = \alpha_i \cdot \mathsf{ewt}(e)$.

Lemma 6.10. *Let $\mathcal{K}$ be a field and let $\mathcal{F}$ be the family of step functions. Then for each weighted timed automaton $\mathcal{A}$ over $\mathcal{K}$, Σ and $\mathcal{F}$ we have*

$$\mathsf{supp}(\|\mathcal{A}\|) = \emptyset \quad \textit{if and only if} \quad \mathsf{supp}(\|R_{\mathsf{step}}(\mathcal{A})\|) = \emptyset.$$

Proof. Let $\mathcal{A} = (\mathcal{L}, \mathcal{C}, E, \mathsf{in}, \mathsf{out}, \mathsf{ewt}, \mathsf{lwt})$ over $\mathcal{K}$, Σ and $\mathcal{F}$. Like the classical region automaton [6], the weighted region automaton $R_{\mathsf{step}}(\mathcal{A})$ is bisimulation equivalent to the infinite state-transition system induced by $\mathcal{A}$. Using this, one can easily show that there is a weight-preserving bijective correspondence between the set of runs of $\mathcal{A}$ and $R_{\mathsf{step}}(\mathcal{A})$, used in the following.

Now, assume $\mathsf{supp}(\|\mathcal{A}\|) \neq \emptyset$. Then there is some $w \in T\Sigma^*$ such that $(\|\mathcal{A}\|, w) \neq 0$. Hence, $\sum\{\mathsf{rwt}(r) : r \text{ is a run of } \mathcal{A} \text{ on } w\} \neq 0$. However, for every run r of $\mathcal{A}$ on w there is a run r' of $R_{\mathsf{step}}(\mathcal{A})$ on $\mathsf{abs}(w)$ such that $\mathsf{rwt}(r') = \mathsf{rwt}(r)$. Moreover, there are no other runs of $R_{\mathsf{step}}(\mathcal{A})$ on $\mathsf{abs}(w)$. Hence, we have $(\|R_{\mathsf{step}}(\mathcal{A})\|, \mathsf{abs}(w)) \neq 0$ and this implies $\mathsf{supp}(\|R_{\mathsf{step}}(\mathcal{A})\|) \neq \emptyset$.

For the other direction, assume $\mathsf{supp}(\|R_{\mathsf{step}}(\mathcal{A})\|) \neq \emptyset$. Hence there is some $v \in (\mathcal{I} \times \Sigma)^*$ such that $(\|R_{\mathsf{step}}(\mathcal{A})\|, v) \neq 0$, i.e., $\sum\{\mathsf{rwt}(r') : r' \text{ is a run of } R_{\mathsf{step}}(\mathcal{A}) \text{ on } v\} \neq 0$. By the definition of $R_{\mathsf{step}}(\mathcal{A})$, there must be some $w \in T\Sigma^*$ such that $v = \mathsf{abs}(w)$. For every run of $R_{\mathsf{step}}(\mathcal{A})$ on v there is a run of $\mathcal{A}$ on w. Moreover, there are no other runs of $\mathcal{A}$ on w. Hence, $\sum\{\mathsf{rwt}(r) : r \text{ is a run of } \mathcal{A} \text{ on } w\} \neq 0$ and we have $(\|\mathcal{A}\|, w) \neq 0$. This implies $\mathsf{supp}(\|\mathcal{A}\|) \neq \emptyset$. ■

Next, we want to go a step further and consider weighted timed automata over fields and more interesting families of weight functions, e.g. linear functions. Clearly, in a construction as above, the running weights of the runs in $\mathcal{A}$ and the corresponding weighted region automaton cannot be equal, since in the weighted region automaton, we abstract from the concrete time delays, which determine the weight of a timed transition of $\mathcal{A}$. However, similarly to the case of semirings which are not zero-divisor free in the previous section, for certain settings we do not need to consider the *exact* weight of a timed transition in order to decide whether the running weight of a run equals 0 or not. We start with the definition of another weighted version of the region automaton.

Let $\mathcal{K}$ be the semiring of the real numbers with addition and multiplication, and $\mathcal{F}$ be the family of linear functions. Given a weighted timed automaton $\mathcal{A} = (\mathcal{L}, \mathcal{C}, E, \mathsf{in}, \mathsf{out}, \mathsf{ewt}, \mathsf{lwt})$ over $\mathcal{K}$, Σ and $\mathcal{F}$, we define the weighted finite automaton $R_{\mathsf{lin}}(\mathcal{A}) = (Q, \Delta, \mathsf{in}', \mathsf{out}', \mathsf{wt})$ over $\mathcal{K}$ and $\mathcal{I} \times \Sigma$, where $(Q, \Delta, \mathsf{in}', \mathsf{out}')$ is the region automaton as defined in Sect. 2.1, and wt is defined as follows: let $t = ((l, r), (I, a), (l', r')) \in \Delta$ stem from e and δ for some $e \in E$ and some $\delta \in I$. Further assume that $\mathsf{lwt}(l)$ is of the form $k \cdot \delta'$ for some $k \in K$ and each $\delta' \in \mathbb{R}_{\geq 0}$. Then, we define $\mathsf{wt}(t) = k \cdot \mathsf{ewt}(e)$ if $\delta \neq 0$, and $\mathsf{wt}(t) = 0$ otherwise.

Lemma 6.11. *Let $\mathcal{K}$ be the semiring of the real numbers with addition and multiplication and $\mathcal{F}$ be the family of linear functions. Then, for each weighted timed automaton $\mathcal{A}$ over $\mathcal{K}$, Σ and $\mathcal{F}$, we have*

$$\mathsf{supp}(\|\mathcal{A}\|) = \emptyset \quad \textit{if and only if} \quad \mathsf{supp}(\|R_{\mathsf{lin}}(\mathcal{A})\|) = \emptyset.$$

PROOF. Let $\mathcal{A} = (\mathcal{L}, \mathcal{C}, E, \mathsf{in}, \mathsf{out}, \mathsf{ewt}, \mathsf{lwt})$ be a weighted timed automaton over $\mathcal{K}$, Σ and $\mathcal{F}$ and let $w \in T\Sigma^*$. Note that for each run of $\mathcal{A}$ on w of the form $r = (l_0, \nu_0) \xrightarrow{\delta_1} \xrightarrow{e_1} \dots \xrightarrow{\delta_{|w|}} \xrightarrow{e_{|w|}} (l_{|w|}, \nu_{|w|})$, by commutativity of $\cdot$ we have

$$\mathsf{rwt}(r) = \mathsf{in}(l_0) \cdot \left(\prod_{1 \leq i \leq |w|} k_{l_{i-1}} \cdot \mathsf{ewt}(e_i) \right) \cdot \mathsf{out}(l_{|w|}) \cdot \prod_{1 \leq i \leq |w|} \delta_i .$$

Let n be the finite number of runs of $\mathcal{A}$ on w and use r^i to denote a run of $\mathcal{A}$ on w for each $i \in \{1, ..., n\}$. By distributivity, we have

$$\begin{aligned}
(\|\mathcal{A}\|, w) &= \sum_{1 \leq i \leq n} \mathsf{rwt}(r^i) \\
&= \sum_{1 \leq i \leq n} \left[\left(\mathsf{in}(l_0^i) \cdot \left(\prod_{1 \leq j \leq |w|} k_{l_{j-1}^i} \cdot \mathsf{ewt}(e_j^i) \right) \cdot \mathsf{out}(l_{|w|}^i) \right) \cdot \prod_{1 \leq j \leq |w|} \delta_j \right] \\
&= \left(\sum_{1 \leq i \leq n} \mathsf{in}(l_0^i) \cdot \left(\prod_{1 \leq j \leq |w|} k_{l_{j-1}^i} \cdot \mathsf{ewt}(e_j^i) \right) \cdot \mathsf{out}(l_{|w|}^i) \right) \cdot \prod_{1 \leq j \leq |w|} \delta_j . \quad (6.1)
\end{aligned}$$

The crucial point here is that the time delays δ_j are the same for each run and thus can be distributed.

Now, assume $\mathsf{supp}(\|\mathcal{A}\|) \neq \emptyset$. Thus, there is some $w \in T\Sigma^*$ such that $(\|\mathcal{A}\|, w) \neq 0$. Hence, the right hand side of equation (6.1) must be different from 0. Since 0 is absorbing, we conclude $\sum_{1 \leq i \leq n} \mathsf{in}(l_0^i) \cdot \left(\prod_{1 \leq j \leq |w|} k_{l_{j-1}^i} \cdot \mathsf{ewt}(e_j^i)\right) \cdot \mathsf{out}(l_{|w|}^i) \neq 0$ and $\prod_{1 \leq j \leq |w|} \delta_j \neq 0$. In particular, $\delta_j \neq 0$ for each $j \in \mathsf{dom}(w)$. However, $\sum_{1 \leq i \leq n} \mathsf{in}(l_0^i) \cdot \left(\prod_{1 \leq j \leq |w|} k_{l_{j-1}^i} \cdot \mathsf{ewt}(e_j^i)\right) \cdot \mathsf{out}(l_{|w|}^i)$ corresponds to the behaviour of $R_{\mathsf{lin}}(\mathcal{A})$ on $\mathsf{abs}(w)$ and thus, this and the definition of wt implies

$$(\|R_{\mathsf{lin}}(\mathcal{A})\|, \mathsf{abs}(w)) = \sum_{1 \leq i \leq n} \mathsf{in}(l_0^i) \cdot \left(\prod_{1 \leq j \leq |w|} k_{l_{j-1}^i} \cdot \mathsf{ewt}(e_j^i)\right) \cdot \mathsf{out}(l_{|w|}^i) \neq 0.$$

Thus, $\mathsf{supp}(\|R_{\mathsf{lin}}(\mathcal{A})\|) \neq \emptyset$.

Now, assume $\mathsf{supp}(\|R_{\mathsf{lin}}(\mathcal{A})\|) \neq \emptyset$. Hence, there is some $v \in (\mathcal{I} \times \Sigma)^*$ such that $(\|R_{\mathsf{lin}}(\mathcal{A})\|, v) \neq 0$. Let $v = (I_1, a_1)(I_2, a_2)...(I_k, a_k)$. We further let n be the number of runs of $R_{\mathsf{lin}}(\mathcal{A})$ on v and use r^i to denote a run of $R_{\mathsf{lin}}(\mathcal{A})$ on v.

$$\begin{aligned} & (\|R_{\mathsf{lin}}(\mathcal{A})\|, v) \\ = & \sum_{1 \leq i \leq n} \{\mathsf{rwt}(r^i) : r^i \text{ is a run of } R_{\mathsf{lin}}(\mathcal{A}) \text{ on } v\} \\ = & \sum_{1 \leq i \leq n} \mathsf{in}((l_0^i, r_0^i)) \cdot \left(\prod_{1 \leq j \leq k} \mathsf{wt}((l_{j-1}^i, r_{j-1}^i), (I_j, a_j), (l_j^i, r_j^i))\right) \cdot \mathsf{out}((l_k^i, r_k^i)) \end{aligned}$$

Now, assume there is some $j \in \{1, ..., k\}$ such that $I_j = [0, 0]$. Then, for each $i \in \{1, ..., n\}$ the transition $t_j = \left((l_{j-1}^i, r_{j-1}^i), (I_j, a_j), (l_j^i, r_j^i)\right)$ stems from e and $\delta = 0$ for some $e \in E$. In this case we have $\mathsf{wt}(t_j) = 0$ by definition. This implies $\mathsf{rwt}(r^i) = 0$ for each $i \in \{1, ..., n\}$ and thus $(\|R_{\mathsf{lin}}(\mathcal{A})\|, v) = 0$, a contradiction. Hence, for each $j \in \{1, ..., k\}$ we must have $I_j \neq [0, 0]$. By the definition of wt we obtain

$$(\|R_{\mathsf{lin}}(\mathcal{A})\|, v) \quad = \quad \sum_{1 \leq i \leq n} \mathsf{in}(l_0^i) \left(\cdot \prod_{1 \leq j \leq k} k_{l_{j-1}^i} \cdot \mathsf{ewt}(e_j^i)\right) \cdot \mathsf{out}(l_k^i)$$

Now, let $w \in \mathsf{abs}^{-1}(v)$ be a timed word. For each run r of $R_{\mathsf{lin}}(\mathcal{A})$ on v there is a corresponding run r' of $\mathcal{A}$ on w. Moreover, there are no other runs of $\mathcal{A}$ on w. As shown in the proof for the other direction, we have

$$(\|\mathcal{A}\|, w) \quad = \quad \left(\sum_{1 \leq i \leq n} \mathsf{in}(l_0^i) \left(\cdot \prod_{1 \leq j \leq |w|} k_{l_{j-1}^i} \cdot \mathsf{ewt}(e_j^i)\right) \cdot \mathsf{out}(l_{|w|}^i)\right) \cdot \prod_{1 \leq j \leq |w|} \delta_j$$

Since for each $j \in \{1, ..., k\}$ we have $I_j \neq [0, 0]$, we also know that $\delta_j \neq 0$ for each $j \in \{1, ..., k\}$. Hence, $\prod_{1 \leq j \leq |w|} \delta_j \neq 0$. However, by assumption we also have $(\|R_{\mathsf{lin}}(\mathcal{A})\|, v) \neq 0$. Hence, $(\|\mathcal{A}\|, w) \neq 0$, and thus $\mathsf{supp}(\|\mathcal{A}\|) \neq \emptyset$. ■

Now, we use the previous two results together with Theorem 2.11 (decidability of the empty support problem for WFA-recognizable series over fields) to obtain the following theorem.

Theorem 6.12. *The emptiness of the support of an $\mathcal{F}$-recognizable timed series over a semiring and a family as specified in Lemmas 6.10 and 6.11 is decidable.*

In the following, we present an interesting result concerning the equality of two $\mathcal{F}$-recognizable timed series over fields. A corresponding result is known for the class of WFA-recognizable series (see e.g. [58]).

Corollary 6.13. *The equality of two given $\mathcal{F}$-recognizable timed series over a semiring and a family as specified in Lemmas 6.10 and 6.11 is decidable.*

PROOF. Let $\mathcal{A}_i = (\mathcal{L}_i, \mathcal{C}_i, E_i, \mathsf{in}_i, \mathsf{out}_i, \mathsf{ewt}_i, \mathsf{lwt}_i)$ be two weighted timed automata over $\mathcal{K}$, Σ and $\mathcal{F}$ such that $\|\mathcal{A}_i\| = \mathcal{T}_i$ (for $i = 1, 2$). We may assume that $\mathcal{L}_1 \cap \mathcal{L}_2 = \emptyset$ and $\mathcal{C}_1 \cap \mathcal{C}_2 = \emptyset$. We define $\mathcal{A} = (\mathcal{L}, \mathcal{C}, E, \mathsf{in}, \mathsf{out}, \mathsf{ewt}, \mathsf{lwt})$, where

- $\mathcal{L} = \mathcal{L}_1 \cup \mathcal{L}_2$, $\mathcal{C} = \mathcal{C}_1 \cup \mathcal{C}_2$, $E = E_1 \cup E_2$,
- $\mathsf{in}(l) = \begin{cases} \mathsf{in}_1(l) & \text{if } l \in \mathcal{L}_1 \\ -\mathsf{in}_2(l) & \text{if } l \in \mathcal{L}_2 \text{ and } \mathsf{in}_2(l) \neq 0 \\ \mathsf{in}_2(l) & \text{otherwise} \end{cases}$
- $\mathsf{out} = \mathsf{out}_1 \cup \mathsf{out}_2$, $\mathsf{ewt} = \mathsf{ewt}_1 \cup \mathsf{ewt}_2$, $\mathsf{lwt} = \mathsf{lwt}_1 \cup \mathsf{lwt}_2$.

Clearly, $\|\mathcal{A}\| = \|\mathcal{A}_1\| - \|\mathcal{A}_2\|$. We further have $\mathcal{T}_1 = \mathcal{T}_2$ if and only if $\mathsf{supp}(\mathcal{T}_1 - \mathcal{T}_2) = \emptyset$. From this together with Theorem 6.12, we can decide whether $\mathcal{T}_1 = \mathcal{T}_2$. ■

In contrast to this, recall that for TA-recognizable timed languages (or, equivalently, $\mathcal{F}$-recognizable timed series over the Boolean semiring), the equality problem is undecidable [6].

Remark 6.14. As mentioned in Sect. 2.3, for the class of WFA-recognizable series over fields and Σ it is not decidable whether the support equals Σ^* [19]. Hence, we cannot use a reduction as above to solve the weighted version of the classical universality problem, which we call the *universal support problem*, for timed series that are recognizable by single-clock weighted timed automata.

6.4 Timed Cut Languages

In this section, we are interested in timed languages consisting of those timed words whose weights under a given $\mathcal{F}$-recognizable timed series exactly correspond to a given value, or whose weights do (not) exceed a given value from the semiring. Sets of words of the second category are known as *cut languages* and play an important role in the theory of weighted finite automata and series (see e.g. [61, 96]). Here, we want to investigate the following problems: for each semiring $\mathcal{K}$ and family $\mathcal{F}$, given an $\mathcal{F}$-recognizable timed series $\mathcal{T}$ over $\mathcal{K}$, $k \in K$ and $\sim \in \{<, \leq, =, \geq, >\}$, is the set $\mathcal{T}_{\sim}^{-1}(k) = \{w \in T\Sigma^* : (\mathcal{T}, w) \sim k\}$ TA-recognizable? We call timed languages of the form $\mathcal{T}_{\sim}^{-1}(k)$ *timed cut languages.* Timed cut languages may give rise to interesting new applications in the analysis of real-time systems. For instance, we may be given a real-time system that consumes a resource, e.g. energy, money or bandwith. While running the system, we want to consume as little as possible of this resource, but nonetheless some minimal conditions on the behaviour of the real-time system, given by a formal specification, must be satisfied. We want to find out the minimal value $k \in K$ such that the system both satisfies the specification and does not consume more than k units of the resource. A first approach may be to check whether $\mathcal{T}_{\leq}^{-1}(k)$ is TA-recognizable for some acceptable and realistic bound $k \in K$, and, if so, to further use model checking techniques to test whether this set satisfies the specification. If yes, we may decrease the bound, otherwise, we may increase it.

Besides new applications in the analysis of real-time systems, the problem of TA-recognizability of timed cut languages shows also strong relations to other problems in the theory of timed languages, as the following example (adopted from [61]) shows: assume there was some semiring $\mathcal{K}$ and a family $\mathcal{F}$ containing $\mathbb{1}$, and we showed that $\mathcal{T}_{=}^{-1}(k)$ is TA-recognizable over Σ for every $\mathcal{F}$-recognizable timed series $\mathcal{T}$ and every $k \in K$. Then, we could conclude that the complement $\bar{L}$ of every *unambiguously* TA-recognizable timed language L was TA-recognizable. The proof of this is as follows: let $L \subseteq T\Sigma^*$ be an unambiguously TA-recognizable timed language over Σ and $\mathcal{A} = (\mathcal{L}, \mathcal{L}_0, \mathcal{L}_f, \mathcal{C}, E)$ be an unambiguous timed automaton over Σ such that $L(\mathcal{A}) = L$. Define $\mathsf{in}(l) = 1$ if $l \in \mathcal{L}_0$ and $\mathsf{in}(l) = 0$ otherwise, $\mathsf{out}(l) = 1$ if $l \in \mathcal{L}_f$ and $\mathsf{out}(l) = 0$ otherwise, and $\mathsf{lwt}(l) = \mathbb{1}$ for each $l \in \mathcal{L}$ and $\mathsf{ewt}(e) = 1$ for each $e \in E$. Then, the behaviour $\|\mathcal{A}'\|$ of the weighted timed automaton $\mathcal{A}' = (\mathcal{L}, \mathcal{C}, E, \mathsf{in}, \mathsf{out}, \mathsf{ewt}, \mathsf{lwt})$ over $\mathcal{K}$, Σ and $\mathcal{F}$ corresponds to the characteristic series 1_L of L. But then by assumption the timed language $\|\mathcal{A}'\|_{=}^{-1}(0) = \bar{L}$ must be TA-recognizable over Σ. However, recall that by now it is not known whether negation preserves TA-recognizability of unambiguous timed languages [106].

Proposition 6.15. *Let $\mathcal{K}$ have characteristic zero and assume that $\mathbb{1} \in \mathcal{F}$. Then, for each TA-recognizable timed language $L \subseteq T\Sigma^*$, there is some $\mathcal{F}$-recognizable timed series $\mathcal{T}$ over $\mathcal{K}$ and Σ such that* $\mathsf{supp}(\mathcal{T}) = L$.

Proof. Let $L \subseteq T\Sigma^*$ be a TA-recognizable timed language over Σ. Then there is a timed automaton $\mathcal{A} = (\mathcal{L}, \mathcal{L}_0, \mathcal{L}_f, \mathcal{C}, E)$ over Σ such that $L(\mathcal{A}) = L$. For each $l \in \mathcal{L}$, we define $\mathsf{in}(l) = 1$ if $l \in \mathcal{L}_0$ and $\mathsf{in}(l) = 0$ otherwise, $\mathsf{out}(l) = 1$ if $l \in \mathcal{L}_f$ and $\mathsf{out}(l) = 0$ otherwise, and we further put $\mathsf{lwt}(l) = \mathbb{1}$. For each $e \in E$, we define $\mathsf{ewt}(e) = 1$. We let $\mathcal{A}' = (\mathcal{L}, \mathcal{C}, E, \mathsf{in}, \mathsf{out}, \mathsf{ewt}, \mathsf{lwt})$ be a weighted timed automaton over $\mathcal{K}$, Σ and $\mathcal{F}$ and show that $\mathsf{supp}(\|\mathcal{A}'\|) = L(\mathcal{A})$.

Let $w \in \mathsf{supp}(\|\mathcal{A}\|)$. Hence, $(\|\mathcal{A}'\|, w) \neq 0$. Now, we know that the running weight of each run of $\mathcal{A}$ on w must be 1 or 0 due to the definition of the weight functions. We can conclude that there must be at least one run of $\mathcal{A}'$ on w with a running weight of 1. Clearly, by definition of the weight function, this run must also be a successful run of $\mathcal{A}$ on w, and thus $w \in L(\mathcal{A})$.

Now, let $w \in L(\mathcal{A})$. Every successful run r of $\mathcal{A}$ on w is also a run of $\mathcal{A}'$ on w with $\mathsf{rwt}(r) = 1$. Moreover, the running weight of each other run of $\mathcal{A}'$ on w must be 0. Since $\mathcal{K}$ has characteristic zero, we can conclude $(\|\mathcal{A}'\|, w) \neq 0$. Hence, $w \in \mathsf{supp}(\|\mathcal{A}'\|)$.

■

As a direct consequence of Prop. 6.15 and non-closure of TA-recognizable timed languages under complement, we get the following lemma.

Proposition 6.16. *Let $\mathcal{K}$ have characteristic zero and assume that $\mathbb{1} \in \mathcal{F}$. Then there is some $\mathcal{F}$-recognizable $\mathcal{T} : T\Sigma^* \to K$ such that $\mathcal{T}_=^{-1}(0)$ is not TA-recognizable over Σ.*

Proof. Let $L \subseteq T\Sigma^*$ such that L is TA-recognizable over Σ and the complement of L, denoted by $\bar{L}$, is not TA-recognizable over Σ. For instance, let L be the language L_{nd} from Ex. 2.3. By Prop. 6.15, L is the support of some $\mathcal{F}$-recognizable timed series $\mathcal{T} : T\Sigma^* \to K$. Hence, $\mathcal{T}_=^{-1}(0) = \bar{L}$ and thus $\mathcal{T}_=^{-1}(0)$ is not TA-recognizable over Σ.

■

The condition on the *semiring* in the two previous propositions is satisfied by every semiring which is not a ring, as e.g. the semiring of the *positive* reals together with addition and multiplication or the min-plus-semiring; but also e.g. by the semiring of the integers with addition and multiplication (which is a ring). In particular, given an $\mathcal{F}$-recognizable timed series $\mathcal{T}$ over the min-plus-semiring, where $\mathcal{F}$ is the family of linear functions, in general we can conlude that there is some $\mathcal{F}$-recognizable timed series $\mathcal{T}$ such that $\mathcal{T}_=^{-1}(0)$ is not TA-recognizable over Σ. However, in contrast to this, Prop. 6.16 cannot be applied if $\mathcal{T}$ is $\mathcal{F}$-recognizable over the semiring of the positive reals with addition and multiplication and the family $\mathcal{F}$ of linear functions. This is because for this setting there is no $\mathbb{1} \in \mathcal{F}$ mapping each time delay to 1. It is an open question whether there is an $\mathcal{F}$-recognizable timed series $\mathcal{T}$ over the semiring of the positive reals and family $\mathcal{F}$ of linear functions such that $\mathcal{T}_=^{-1}(0)$ is not TA-recognizable. Notice that for the family of step functions we can apply Prop. 6.16, since $\mathbb{1}$ is simply the constant function mapping each time delay to 1. This raises the interesting question whether in

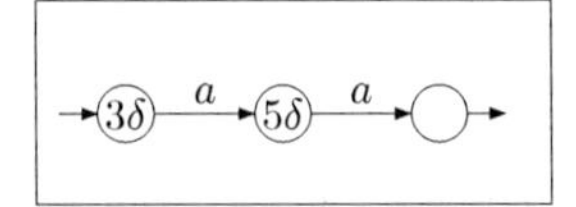

Figure 6.1: A weighted timed automaton $\mathcal{A}$ with $\|\mathcal{A}\|_{=}^{-1}(6)$ not TA-recognizable

general, assuming that $\mathcal{K}$ is fixed, we cannot conclude from a "negative" result for the family of step functions, that a negative result also holds for the (more expressive and harder) family of linear functions.

The negative result in Prop. 6.16 mainly relies on the non-closure of TA-recognizable timed languages under complement. The question arises whether we can obtain a positive result for $k \neq 0$. Unfortunately, for the most interesting class of weighted timed automata over the reals and linear functions we have to give a negative answer, even if we confine our study to *unambiguously* $\mathcal{F}$-recognizable timed series.

Proposition 6.17. *Let $\mathcal{K}$ be one of the following semirings*

- *the min-plus-semiring,*
- *the max-plus-semiring,*
- *the positive real numbers with addition and multiplication,*

and $\mathcal{F}$ be the family of linear functions. Then there is some unambiguously $\mathcal{F}$-recognizable $\mathcal{T} : T\Sigma^* \to K$ and some $k \in K\backslash\{0\}$ such that $\mathcal{T}_{=}^{-1}(k)$ is not TA-recognizable over Σ.

Proof. We show the proof for the min-plus-semiring. Let $\mathcal{A}$ be the weighted timed automaton over the min-plus-semiring and the family of linear functions shown in Fig. 6.1. We further let $k = 6$. Now assume there is a timed automaton $\mathcal{A}'$ such that $L(\mathcal{A}') = \|\mathcal{A}\|_{=}^{-1}(k)$. Then there must be a successful run of $\mathcal{A}'$ on $(a, 1.5)(a, 1.8)$. Then, since there are only finitely many edges in $\mathcal{A}'$ and we are only allowed to check clock variables against natural numbers, we can conclude that there is some $\epsilon \in \mathbb{R}_{\geq 0}$ such that the sequence of edges of each run of $\mathcal{A}'$ on w is also the basis for a run of $\mathcal{A}'$ on $(a, 1.5 + \epsilon)(a, 1.8)$, a contradiction to $L(\mathcal{A}') = \|\mathcal{A}\|_{=}^{-1}(k)$. ∎

Next, we show that if we let $\mathcal{F}$ be the family of *step functions* and consider unambiguously $\mathcal{F}$-recognizable timed series, we can give a positive answer. This may be not too surprising, as in this case the number of weights occuring in runs of the weighted timed automaton is finite and we can apply proof methods known from the theory of weighted finite automata (see e.g. [19]). For the sake of completeness we present some of the results in the following. They may also serve as a starting point for further research on other kinds of weight functions. First, let $\mathcal{F}$ be a family of step functions from $\mathbb{R}_{\geq 0}$ to K and $\eta : K \to K'$ be a semiring morphism. Then, we define $\eta(\mathcal{F})$ to be the family of step functions from $\mathbb{R}_{\geq 0}$ to K' that is obtained from the functions in $\mathcal{F}$ by applying η to each coefficient $\alpha \in K$.

Lemma 6.18. *Let $\mathcal{K}, \mathcal{K}'$ be two semirings and let $\mathcal{F}$ be the family of step functions. If $\mathcal{T} : T\Sigma^* \to K$ is an $\mathcal{F}$-recognizable timed series over $\mathcal{K}$ and $\eta : K \to K'$ is a semiring morphism, then $\eta \circ \mathcal{T}$ is $\eta(\mathcal{F})$-recognizable over $\mathcal{K}'$.*

Proof. Let $\mathcal{T} : T\Sigma^* \to K$ be an $\mathcal{F}$-recognizable timed series over $\mathcal{K}$. Then there is a weighted timed automaton $\mathcal{A}$ over $\mathcal{K}$, Σ and $\mathcal{F}$ such that $\|\mathcal{A}\| = \mathcal{T}$. We obtain the weighted timed automaton $\mathcal{A}'$ over $\mathcal{K}'$, Σ and $\mathcal{F}$ by replacing all coefficients $k \in K$ occuring in the weight functions of $\mathcal{A}$ by $\eta(k) \in K'$. Then one can show that there is a bijective correspondence between the set of runs of $\mathcal{A}$ and the set of runs of $\mathcal{A}'$. Since η is a morphism, we can moreover prove $\mathsf{rwt}(r') = \eta(\mathsf{rwt}(r))$, where r is a run of $\mathcal{A}$, and r' is the corresponding run of $\mathcal{A}'$. Finally, we have for each $w \in T\Sigma^*$

$$\begin{aligned}(\|\mathcal{A}'\|, w) &= \sum\{\mathsf{rwt}(r') | r' \text{ is a run of } \mathcal{A}' \text{ on } w\} \\ &= \sum\{\eta(\mathsf{rwt}(r)) | r \text{ is a run of } \mathcal{A} \text{ on } w\} \\ &= \eta(\sum\{rwt(r) | r \text{ is a run of } \mathcal{A} \text{ on } w\}) \\ &= \eta(\|\mathcal{A}\|, w).\end{aligned}$$

■

Before we come to the next result, we introduce some auxiliary notion. Let I be an interval over $\mathbb{R}_{\geq 0}$ with borders in $\mathbb{N}$ and x be a clock variable. Then, we use $\mathsf{cc}(I, x)$ to denote the clock constraint defined by

$$\mathsf{cc}(I, x) = \begin{cases} a < x \wedge x < b & \text{if } I = (a, b), \\ a \leq x \wedge x < b & \text{if } I = [a, b), \\ a < x \wedge x \leq b & \text{if } I = (a, b], \\ a \leq x \wedge x \leq b & \text{if } I = [a, b]. \end{cases}$$

Proposition 6.19. *Let $\mathcal{K}$ be one of the following semirings*

1. *the min-plus-semiring,*
2. *the Viterbi-semiring,*
3. $([1, \infty), \min, \cdot, \infty, 1)$,

and let $\mathcal{F}$ be the family of step functions. Then for each $k \in K\backslash\{0\}$ and each unambiguously $\mathcal{F}$-recognizable timed series over $\mathcal{K}$, the timed language $\mathcal{T}_{=}^{-1}(k)$ is unambiguously TA-recognizable.

PROOF. 1. Let $\mathcal{K}$ be the min-plus-semiring, $\mathcal{F}$ be the family of step functions, $k \in \mathbb{R}_{\geq 0}$ and $\mathcal{T} : T\Sigma^* \rightarrow K$ be an unambiguously $\mathcal{F}$-recognizable timed series. Then there is an unambiguous weighted timed automaton $\bar{\mathcal{A}}$ over $\mathcal{K}$, Σ and $\mathcal{F}$ such that $\|\bar{\mathcal{A}}\| = \mathcal{T}$. Let $\mathcal{A}$ be the final-location-normalized weighted timed automaton obtained from $\bar{\mathcal{A}}$ by applying the algorithm given in Lemma 4.9. Note that $\mathcal{A}$ is still unambiguous and we have $(\|\mathcal{A}\|, w) = (\|\bar{\mathcal{A}}\|, w)$ for each $w \in T\Sigma^+$. We will construct an unambiguous timed automaton $\mathcal{A}'$ over Σ such that $L(\mathcal{A}') = \{w \in T\Sigma^+ : (\mathcal{T}, w) = k\}$. We summarize the idea of the construction. Assume that N is the minimal value in K occuring as a weight in $\mathcal{A}$. Then we only have to consider runs of length smaller than or equal to the ceiling $\lceil \frac{k}{N} \rceil$ of $\frac{k}{N}$. This is because every other run necessarily has a running weight greater than k. Thus, the set of occuring weights relevant for solving the problem is finite and can be remembered within the finite control part of a timed automaton.

Let $\mathcal{A} = (\mathcal{L}, \mathcal{C}, E, \mathsf{in}, \mathsf{out}, \mathsf{ewt}, \mathsf{lwt})$. We use l_f to denote the single sink of $\mathcal{A}$. For each $l \in \mathcal{L}$, we assume $\mathsf{lwt}(l)$ to be of the form $\min\{\alpha_i^l + \chi_{A_i^l} | 1 \leq i \leq n_l\}$ for some $n_l \in \mathbb{N}$, $\alpha_i^l \in \mathbb{R}_{\geq 0} \cup \{\infty\}$, intervals A_i^l over $\mathbb{R}_{\geq 0}$ with borders in $\mathbb{N}$ such that $A_j^l \cap A_k^l = \emptyset$ for $j, k \in \{1, ..., n_l\}$ such that $j \neq k$ and $\bigcup_{1 \leq i \leq n_l} A_i^l = \mathbb{R}_{\geq 0}$. Define $N = \min(\mathsf{wgt}(\mathcal{A})\backslash\{0\})$. We define the timed automaton $\mathcal{A}' = (\mathcal{L}', \mathcal{L}'_0, \mathcal{L}'_f, \mathcal{C}', E')$ over Σ by

- $\mathcal{L}' = \mathcal{L} \times \{m \in K : m = m_1 + ... + m_p, m_1, ..., m_p \in \mathsf{wgt}(\mathcal{A}), p \in \{1, ..., \lceil \frac{k}{N} \rceil\}\}$,
- $\mathcal{L}'_0 = \{(l, k') : l \in \mathcal{L}$ such that $\mathsf{in}(l) \neq \infty, k' = \mathsf{in}(l)\}$,
- $\mathcal{L}'_f = \{(l_f, k)\}$,
- $\mathcal{C}' = \mathcal{C} \,\dot\cup\, \{x\}$,
- $E' = \{((l, m), a, \phi', \lambda', (l', m')) : (l, a, \phi, \lambda, l') \in E$ such that $\phi' = \phi \wedge \mathsf{cc}(A_i^l, x), \lambda' = \lambda \cup \{x\}, m' = m + \alpha_i^l + \mathsf{ewt}(l, a, \phi, \lambda, l')$ for every $i \in \{1, ..., n_l\}\}$. We say that $((l, m), a, \phi', \lambda', (l', m'))$ stems from $(l, a, \phi, \lambda, l')$.

We show that $L(\mathcal{A}') = \|\mathcal{A}\|_{=}^{-1}(k)$. Let $w \in T\Sigma^+$. First, assume $w \in L(\mathcal{A}')$. Then there is a successful run $((l_0, m_0), \nu_0) \xrightarrow{\delta_1} \xrightarrow{e'_1} ((l_1, m_1), \nu_1) \xrightarrow{\delta_2} \xrightarrow{e'_2} ... \xrightarrow{\delta_{|w|}} \xrightarrow{e'_{|w|}} ((l_{|w|}, k_{|w|}), \nu_{|w|})$ of $\mathcal{A}'$ on w, where $l_{|w|} = l_f$ and $k_{|w|} = k$. For every $i \in \mathsf{dom}(w)$, let $e_i \in E$ such that e'_i stems from e_i. Consider the run $r = (l_0, \nu_{0|\mathcal{C}}) \xrightarrow{\delta_1} \xrightarrow{e_1} (l_1, \nu_{1|\mathcal{C}}) \xrightarrow{\delta_2} \xrightarrow{e_2} ... \xrightarrow{\delta_{|w|}} \xrightarrow{e_{|w|}} (l_{|w|}, \nu_{|w||\mathcal{C}})$. Clearly this is a run of $\mathcal{A}$ on w. We show $\mathsf{rwt}(r) = k$. For each $i \in \mathsf{dom}(w)$ there is a unique $j \in \{1, ..., n_{l_{i-1}}\}$ such that $\delta_i \in A_j^{l_{i-1}}$, in the following denoted by j_i. For each $i \in \{0, ..., |w|\}$, we define ω_i to be the sum of $\mathsf{in}(l_0), \alpha_{j_1}^{l_0}, \mathsf{ewt}(e_1), ..., \alpha_{j_i}^{l_{i-1}}, \mathsf{ewt}(e_i)$. We show inductively that $m_i = \omega_i$ for each $i \in \{0, ..., |w|\}$. For the base case, $p = 0$, this is trivially the case. So assume that $m_i = \omega_i$ for each $i \in \{0, ..., p\}$ for some $p < |w|$. Notice that by the choice of the reset sets in E', we have $(\nu_p + \delta_{p+1})(x) = \delta_{p+1}$. Since $(\nu_p + \delta_{p+1}) \models \phi'_{p+1}$, we can conclude that $\phi' = \phi_{p+1} \wedge \mathsf{cc}(A^{l_p}_{j_{p+1}}, x)$. But this implies $m_{p+1} = m_p + \alpha^{l_p}_{j_{p+1}} + \mathsf{ewt}(e_{p+1}) = \omega_p + \alpha^{l_p}_{j_{p+1}} + \mathsf{ewt}(e_{p+1}) = \omega_{p+1}$. Hence, we also have $\omega_{|w|} = k$. However, we also have $\mathsf{rwt}(r) = \omega_{|w|} = k$ (since $\mathsf{out}(l_f) = 0$). From $\mathcal{A}$ being unambiguous, it follows $(\|\mathcal{A}\|, w) = k$.

Now, let $w \in \|\mathcal{A}\|^{-1}_{=}(k)$. Since $\mathcal{A}$ is unambiguous, there is exactly one run r of $\mathcal{A}$ on w with $\mathsf{rwt}(r) = k$, the running weights of all the other runs of $\mathcal{A}$ on w equal 0. So let r be of the form $(l_0, \nu_0) \xrightarrow{\delta_1} \xrightarrow{e_1} ... \xrightarrow{\delta_{|w|}} \xrightarrow{e_{|w|}} (l_{|w|}, \nu_{|w|})$. Let $i \in \mathsf{dom}(w)$. We let j_i be the unique $j \in \{1, ..., n_{l_{i-1}}\}$ such that $\delta_i \in A^{l_{i-1}}_j$. Clearly, we have $\mathsf{lwt}(l_{i-1})(\delta_i) = \alpha^{l_{i-1}}_{j_i}$. We assume e_i to be of the form $(l_{i-1}, a_i, \phi_i, \lambda_i, l_i)$ and define $e'_i = ((l_{i-1}, m_{i-1}), a_i, \phi'_i, \lambda'_i, (l_i, m_i))$ such that $\phi'_i = \phi_i \wedge \mathsf{cc}(A^{l_{i-1}}_{j_i}, y)$, $\lambda'_i = \lambda_i \cup \{x\}$ and $m_i = m_{i-1} + \alpha^{l_{i-1}}_{j_i} + \mathsf{ewt}(e_i)$, where $m_0 = \mathsf{in}(l_0)$. Hence, $e'_i \in E'$. One can easily prove that $((l_0, m_0), \nu'_0) \xrightarrow{\delta_1} \xrightarrow{e'_1} ... \xrightarrow{\delta_{|w|}} \xrightarrow{e'_{|w|}} ((l_{|w|}, m_{|w|}), \nu'_{|w|})$ is a run of $\mathcal{A}'$ on w. Clearly, $m_{|w|} = \mathsf{rwt}(r) = k$ and thus this run is a successful run of $\mathcal{A}'$ on w. Hence, $w \in L(\mathcal{A}')$.

We thus have proved that $\mathcal{T}^{-1}_{=}(k)$ is recognizable by the timed automaton $\mathcal{A}'$ over Σ. Now, if $(\|\bar{\mathcal{A}}\|, \varepsilon) = k$, then define $\mathcal{A}''$ to be the timed automaton over Σ obtained from $\mathcal{A}'$ by adding a new location $l \notin \mathcal{L}'$ such that l both is initial and final. If on the other hand $(\|\bar{\mathcal{A}}\|, \varepsilon) \neq k$, we let $\mathcal{A}'' = \mathcal{A}'$. Then, we have $L(\mathcal{A}'') = (\mathcal{T})^{-1}_{=}(k)$.

2. Now, let $ln(n)$ denote the natural logarithm of a number n. We define the semiring morphism $\eta : [0, 1] \rightarrow \mathbb{R}_{\geq 0} \cup \{\infty\}$ by $\eta(n) = -ln(n)$ for each $n \in [0, 1]$. Let $\mathcal{T}$ be an $\mathcal{F}$-recognizable timed series over the Viterbi semiring, and let $k \in [0, 1] \backslash \{0\}$. Then, by Lemma 6.18, $\eta \circ \mathcal{T}$ is $\eta(\mathcal{F})$-recognizable over the min-plus-semiring. Using 1., we know that $\{w \in T\Sigma^* | (\eta \circ \mathcal{T}, w) = \eta(k)\}$ is TA-recognizable. But this implies that $\{w \in T\Sigma^* | (\mathcal{T}, w) = k\}$ is TA-recognizable.

3. This can be proved similarly to the proof of 2. by using the semiring morphism $\eta' : [1, \infty) \rightarrow \mathbb{R}_{\geq 0} \cup \{\infty\} : n \mapsto ln(n)$. ■

A similar proof method to that of the first claim of Prop. 6.19 can be used to prove the claim for the following semirings:

- max-plus-semiring,
- $([0, 1], \max, \min, 0, 1)$,
- min-max-semiring,
- $(\mathbb{R}_{\geq 0} \cup \{\infty\}, \min, \cdot, \infty, 1)$,
- $([0, 1] \cup \{\infty\}, \min, \cdot, \infty, 1)$.

Using the semiring morphism η from $(\mathbb{R}_{\geq 0} \cup \{\infty\}, \max, \min, 0, \infty)$ to the min-max-semiring as defined in the proof of the second claim of Prop. 6.19, we obtain the claim also for $(\mathbb{R}_{\geq 0} \cup \{\infty\}, \max, \min, 0, \infty)$.

Remark 6.20. Note that the mapping η used in the proof of Prop. 6.19 is also an isomorphism between the more general algebraic structure $([0,1], \max, \cdot, 0, 1, (k^{\delta})_{\delta \in \mathbb{R}_{\geq 0}})$ and $(\mathbb{R}_{\geq 0} \cup \{\infty\}, \max, +, \infty, 0, (k \cdot \delta)_{\delta \in \mathbb{R}_{\geq 0}})$. From this we can conclude that results concerning weighted timed automata over the min-plus-semiring and linear functions presented here or e.g. in [22] also hold for weighted timed automata over the Viterbi semiring and the family of functions of the form $f(\delta) = k^{\delta}$ for each $\delta \in \mathbb{R}_{\geq 0}$ and some $k \in [0,1]$.

Proposition 6.21. *Let $\mathcal{K}$ be such that $(K, \cdot, 1)$ is locally finite, let $\mathcal{F}$ be the family of step functions, and let $k \in K \backslash \{0\}$. Then for each unambiguously $\mathcal{F}$-recognizable timed series $\mathcal{T}$ over $\mathcal{K}$, the timed language $\mathcal{T}_{=}^{-1}(k)$ is unambiguously TA-recognizable.*

Proof. (Sketch) We proceed similarly to the proof of the first claim in Prop. 6.19. The crucial point is that $\mathsf{wgt}(\mathcal{A})$ (for $\mathcal{A}$ being the weighted timed automaton over $\mathcal{K}$, Σ and $\mathcal{F}$ recognizing $\mathcal{T}$) is finite and thus, by assumption, also is the submonoid generated by $\mathsf{wgt}(\mathcal{A})$. Thus, the weights arising in $\mathcal{A}$ while reading a timed word can be remembered in the finite control part of a timed automaton. For this, we introduce a new clock variable used to check whether we have reached the weight k or not. ■

Proposition 6.22. *Let $\mathcal{K}$ be the min-max-semiring, let $\mathcal{F}$ be the family of linear functions of the form $c \cdot \delta$ such that $c \in \mathbb{Z}$, and let $k \in \mathbb{Z}$. Then, for each unambiguously $\mathcal{F}$-recognizable timed series $\mathcal{T}$ over $\mathcal{K}$, the timed language $\mathcal{T}_{=}^{-1}(k)$ is unambiguously TA-recognizable.*

Proof. Let $\mathcal{K}$ be the min-max-semiring, let $\mathcal{F}$ be the family of linear functions of the form $c \cdot \delta$ such that $c \in \mathbb{Z}$, let $k \in \mathbb{Z}$ and let $\mathcal{T} : T\Sigma^* \to K$ be unambiguously $\mathcal{F}$-recognizable. Then there is an unambiguous weighted timed automaton $\mathcal{A} = (\mathcal{L}, \mathcal{C}, E, \mathsf{in}, \mathsf{out}, \mathsf{ewt}, \mathsf{lwt})$ over $\mathcal{K}$, Σ and $\mathcal{F}$ such that $\|\mathcal{A}\| = \mathcal{T}$.

First, assume $k \geq 0$. We define the timed automaton $\mathcal{A}' = (\mathcal{L}', \mathcal{L}'_0, \mathcal{L}'_f, \mathcal{C}', E'_0)$ such that

- $\mathcal{L}' = \mathcal{L} \times \{0,1\}$
- $\mathcal{L}'_0 = (\{l \in \mathcal{L} : \mathsf{in}(l) < k\} \times \{0\}) \cup (\{l \in \mathcal{L} : \mathsf{in}(l) = k\} \times \{1\})$
- $\mathcal{L}'_f = \{l \in \mathcal{L} : \mathsf{out}(l) \leq k\} \times \{1\}$
- $\mathcal{C}' = \mathcal{C} \dot{\cup} \{x\}$,
- $E' = E_1 \cup E_2 \cup E_3 \cup E_4 \cup E_5 \cup E_6 \cup E_7$,

where

- $E_1 = \{((l,0),a,\phi',\lambda',(l',0)) : (l,a,\phi,\lambda,l') \in E$ such that $\mathsf{ewt}((l,a,\phi,\lambda,l')) < k$, $\mathsf{lwt}(l)(\delta) = k' \cdot \delta$ for some $k' > 0$, and $\phi' = \phi \wedge x < \frac{k}{k'}, \lambda' = \lambda \cup \{x\}\}$,
- $E_2 = \{((l,0),a,\phi',\lambda',(l',1)) : (l,a,\phi,\lambda,l') \in E$ such that $\mathsf{ewt}((l,a,\phi,\lambda,l')) = k$, $\mathsf{lwt}(l)(\delta) = k' \cdot \delta$ for some $k' > 0$, and $\phi' = \phi \wedge x \leq \frac{k}{k'}, \lambda' = \lambda \cup \{x\}\}$,
- $E_3 = \{((l,0),a,\phi',\lambda',(l',1)) : (l,a,\phi,\lambda,l') \in E$ such that $\mathsf{ewt}((l,a,\phi,\lambda,l')) \leq k$, $\mathsf{lwt}(l)(\delta) = k' \cdot \delta$ for some $k' > 0$, and $\phi' = \phi \wedge x = \frac{k}{k'}, \lambda' = \lambda \cup \{x\}\}$,
- $E_4 = \{((l,1),a,\phi',\lambda',(l',1)) : (l,a,\phi,\lambda,l') \in E$ such that $\mathsf{ewt}((l,a,\phi,\lambda,l')) \leq k$, $\mathsf{lwt}(l)(\delta) = k' \cdot \delta$ for some $k' > 0$, and $\phi' = \phi \wedge x \leq \frac{k}{k'}, \lambda' = \lambda \cup \{x\}\}$,
- $E_5 = \{((l,0),a,\phi,\lambda',(l',0)) : (l,a,\phi,\lambda,l') \in E$ such that $\mathsf{ewt}((l,a,\phi,\lambda,l')) < k$, $\mathsf{lwt}(l)(\delta) = k' \cdot \delta$ for some $k' \leq 0$, and $\lambda' = \lambda \cup \{x\}\}$,
- $E_6 = \{((l,0),a,\phi,\lambda',(l',1)) : (l,a,\phi,\lambda,l') \in E$ such that $\mathsf{ewt}((l,a,\phi,\lambda,l')) = k$, $\mathsf{lwt}(l)(\delta) = k' \cdot \delta$ for some $k' \leq 0$, and $\lambda' = \lambda \cup \{x\}\}$,
- $E_7 = \{((l,1),a,\phi,\lambda',(l',1)) : (l,a,\phi,\lambda,l') \in E$ such that $\mathsf{ewt}((l,a,\phi,\lambda,l')) \leq k$, $\mathsf{lwt}(l)(\delta) = k' \cdot \delta$ for some $k' \leq 0$, and $\lambda' = \lambda \cup \{x\}\}$.

The idea of this construction can be summarized as follows. In the second component of each location we remember whether we have already reached a weight of k (1) or not (0). The former is the case if either the weight for entering a location, for taking an edge or leaving a location equals k, or if we stay for exactly as many time units in a location l such that the weight of a timed transition in l equals k. For the timed transition, we get this information by a newly introduced clock variable x which measures for each location the time that has been spent in this locations. At every edge a clock constraint controls the behaviour of $\mathcal{A}'$ depending on whether we have already reached k and on how much time we have already spent in the source location. Doing so, we distinguish between positive (E_1, E_2, E_3, E_4) and negative (E_5, E_6, E_7) coefficients used in the location weight functions.

Next, we show $L(\mathcal{A}') = \|\mathcal{A}\|_{=}^{-1}(k)$. For this let $w \in T\Sigma^*$ be of the form $(a_1,t_1)...(a_n,t_n)$.

For the first direction, assume $w \in L(\mathcal{A}')$. Hence there is a successful run $r' = ((l_0,b_0),\nu_0) \xrightarrow{\delta_1}\xrightarrow{e'_1} \ldots \xrightarrow{\delta_n}\xrightarrow{e'_n} ((l_n,b_n),\nu_n)$ of $\mathcal{A}'$ on w. Then, the run $r = (l_0,\nu_{0|C}) \xrightarrow{\delta_1}\xrightarrow{e'_1} \ldots \xrightarrow{\delta_n}\xrightarrow{e'_n} (l_n,\nu_{n|C})$, where e_i is obtained from e'_i by removing all information on x, is a run of $\mathcal{A}$ on w. We show that $\mathsf{rwt}(r) = k$. Since r' is successful, we must have $b_n = 1$ and either $\mathsf{in}(l_0) < k$ and $b_0 = 0$, or $\mathsf{in}(l_0) = k$ and $b_0 = 1$. In the former case, there must be some $i \in \{1,...,n\}$ such that $e_i \in E_2 \cup E_3 \cup E_6$, and for all $j \in \{0,...,i-1\}$, we have $e_j \in E_1 \cup E_5$ and for all $j \in \{i+1,...,n\}$, we have $e_j \in E_4 \cup E_7$.

First, let $j \in \{1, ..., i-1\}$. If $e_j \in E_1$, then $\mathsf{ewt}(e_j) < k$ and $\mathsf{lwt}(l_{j-1})(\delta) = k' \cdot \delta$ for some $k' > 0$. By the fact that x is reset in every edge of $\mathcal{A}'$ and $\nu_{j-1} + \delta_j \models \phi'_j$, we know that $(\nu_{j-1} + \delta_j)(x) = \delta_j$ and thus $\delta_j < \frac{k}{k'}$. Hence, we have $\mathsf{lwt}(l_{j-1})(\delta_j) < k$. If $e_j \in E_5$, we also have $\mathsf{ewt}(e_j) < k$. Moreover, $\mathsf{lwt}(l_{j-1})(\delta) = k' \cdot \delta$ for some $k' \leq 0$. Thus, we obtain $\mathsf{lwt}(l_{j-1})(\delta_j) < k$ anyway. In all these case, we have not reached a weight of k yet.

Second, let $j = i$. If $e_j \in E_2$, then $\mathsf{ewt}(e_j) = k$ and $\mathsf{lwt}(l_{j-1})(\delta_j) \leq k$. If $e_j \in E_3$, then $\mathsf{ewt}(e_j) \leq k$ and $\mathsf{lwt}(l_{j-1})(\delta_j) = k$. If, finally, $e_j \in E_6$, we have $\mathsf{ewt}(e_j) = k$ and $\mathsf{lwt}(l_{j-1})(\delta_j) < k$. In all these cases, we have reached a weight of k.

Third, let $j \in \{i+1, ..., n\}$. If $e_j \in E_4$, then $\mathsf{ewt}(e_j) \leq k$ and $\mathsf{lwt}(l_{j-1})(\delta_j) \leq k$. If $e_j \in E_7$, then $\mathsf{ewt}(e_j) \leq k$ and $\mathsf{lwt}(l_{j-1})(\delta_j) < k$. In all these cases, we do not exceed the weight of k.

Notice that we also have $\mathsf{out}(l_n) \leq k$, and thus we do not exceed the weight k. Hence, we can conclude $\mathsf{rwt}(r) = \max\{\mathsf{lwt}(l_{j-1})(\delta_j), \mathsf{ewt}(e_j) : 1 \leq j \leq n\} = k$, and thus (as $\mathcal{A}$ is unambiguous) $w \in \|\mathcal{A}\|_{=}^{-1}(k)$.

Now, we assume the other case, namely that $\mathsf{in}(l_0) = k$ and $b_0 = 1$. Then, for all $i \in \{1, ..., n\}$, we have $e_i \in E_4 \cup E_7$. Using the same lines of argumentation as above, we conclude that the weights of all transitions do not exceed k. Hence, we have $\mathsf{rwt}(r) = k$, and thus, by unambiguity of $\mathcal{A}'$, $w \in \|\mathcal{A}\|_{=}^{-1}(k)$.

The proof for the other direction can be done analogously. Hence, we have $L(\mathcal{A}') = \|\mathcal{A}'\|_{=}^{-1}(k)$.

For $k < 0$, we can proceed in a similar way. ∎

Remark 6.23. We can use similar proof methods to show the claim of Prop. 6.22 for each semiring $\mathcal{K}$ such that $K \supseteq \mathbb{R}_{\geq 0}$ and $\cdot = \min$ or $\cdot = \max$, and $\mathcal{F}$ the family of linear function.

Next, we can again generalize results from the theory of weighted finite automata and WFA-recognizable series, respectively, to recognizable timed series over the family of step functions owing to the *finite* number of weights occuring in a corresponding weighted timed automaton. Notice that in the next theorem, in opposition to Prop. 6.19, we do not require $\mathcal{T}$ to be unambiguously $\mathcal{F}$-recognizable.

Theorem 6.24. *1. Let $\mathcal{K}$ be one of the following semirings*

a) min-plus-semiring,

b) min-max-semiring,

c) $(\mathbb{R}_{\geq 0} \cup \{\infty\}, \min, \cdot, \infty, 1)$,

d) $([0,1] \cup \{\infty\}, \min, \cdot, \infty, 1)$,

e) $([1, \infty], \min, \cdot, \infty, 1)$.

Let $\mathcal{F}$ be the family of step functions, and $k \in K \backslash \{0\}$. Then, for each $\mathcal{F}$-recognizable timed series $\mathcal{T}$ over $\mathcal{K}$, the timed language $\mathcal{T}_{\leq}^{-1}(k)$ is TA-recognizable.

2. *Let $\mathcal{K}$ be one of the following semirings*

   a) max-plus-semiring,

   b) $(\mathbb{R}_{\geq 0} \cup \{\infty\}, \max, \min, 0, \infty)$,

   c) $([0,1], \max, \min, 0, 1)$,

   d) $([0,1], \max, \cdot, 0, 1)$.

   Let $\mathcal{F}$ be the family of step functions, and let $k \in K \setminus \{0\}$. Then, for each $\mathcal{F}$-recognizable timed series $\mathcal{T}$ over $\mathcal{K}$, the timed language $\mathcal{T}_{\geq}^{-1}(k)$ is TA-recognizable.

PROOF. We proceed as in the proof of Prop. 6.19. More detailed, we may start by giving a direct construction of a timed automaton recognizing the timed cut language of a weighted timed automaton over the min-plus-semiring. This construction goes along the lines of the construction in the proof of Prop. 6.19(a), but note that here we do not need unambiguity due to the relation $\leq$. Then the result follows also for weighted timed automata over the Viterbi semiring using the homomorphism η defined in the proof of Prop. 6.19(b). For this, we use the fact that η is an anti-isomorphism with respect to the natural orderings in the Viterbi semiring and the min-plus-semiring. Thus, we have

$$\{w \in T\Sigma^* | (\eta \circ \mathcal{T}, w) \leq \eta(k)\} = \{w \in T\Sigma^* | (\mathcal{T}, w) \geq k)\}$$

and we conclude that $\{w \in T\Sigma^* | (\mathcal{T}, w) \geq k)\}$ is TA-recognizable over Σ. ■

6.5 Conclusions and Further Research

In this chapter, we investigated decision problems concerning the supports and timed cut languages of recognizable timed series. We believe that this work is only the beginning of fruitful further research within this area, as there are a lot of open problems worth considering. For instance, some results of this chapter do not cover recognizable timed series over certain important semirings and families of functions. In particular, we would like to know whether there is a recognizable timed series over the semiring of the positive reals with addition and multiplication and the family of linear functions such that $\mathcal{T}^{-1}(0)$ is not TA-recognizable. We also note that for some results we assume that $\mathcal{F}$ contains $\mathbb{1}$. However, for some settings of semirings and families, no such function exists. As an example, consider the semiring of the positive reals with addition and multiplication and the family of linear functions. A simple solution for this is to add the constant function mapping each time delay to 1. However, for some results it is not possible to include both linear and constant functions in $\mathcal{F}$. For instance, we cannot apply the proof method of Lemma 6.11 when $\mathcal{F}$ contains both linear and constant functions. For this reason, we want to investigate whether we can extend some of the results to more general families of functions. Of course, one may also think of completely other weight functions, for instance logarithmic or exponential functions, and other semirings.

We further like to mention that the results presented in this chapter may be used to obtain some decidability results for the weighted relative distance logic, particularly since the constructions in Chapter 5 are effective. For instance, for the widely-used setting of the min-plus-semiring and the family of linear functions, we obtain the following consequence using Theorems 6.2 and 2.5.

Corollary 6.25. *Let $\mathcal{K}$ be the min-plus-semiring and let $\mathcal{F}$ be the family of linear functions. Then it is decidable, for a given sentence $\varphi \in \mathsf{sR}\mathcal{L}\overleftarrow{\mathsf{d}}(\mathcal{K}, \Sigma, \mathcal{F})$, whether $\mathsf{supp}([\![\varphi]\!]) = \emptyset$. Hence the satisfiability problem of $\mathsf{sR}\mathcal{L}\overleftarrow{\mathsf{d}}(\mathcal{K}, \Sigma, \mathcal{F})$ for the min-plus-semiring and linear functions is decidable.*

Corollary 6.13 leads to the following consequence.

Corollary 6.26. *Let $\mathcal{K}$ be the semiring over the reals with addition and multiplication and let $\mathcal{F}$ be the family of linear functions. Then it is decidable, for two given sentences $\varphi_1, \varphi_2 \in \mathsf{sR}\mathcal{L}\overleftarrow{\mathsf{d}}^{\mathsf{b}}(\mathcal{K}, \Sigma, \mathcal{F})$, whether $\varphi_1 \equiv \varphi_2$.*

In future research, we would like to investigate other weighted versions of classical decision problems, for instance the language inclusion problem, in which one asks whether for two recognizable timed series $\mathcal{T}_1, \mathcal{T}_2$ over an ordered semiring we have $(\mathcal{T}_1, w) \leq (\mathcal{T}_2, w)$ for all timed words $w \in T\Sigma^*$. Since the language inclusion problem is not decidable in general [6], we may have to confine the research on subclasses of weighted timed automata, e.g. with a single clock variable [91] or weighted event-clock automata [7]. Also, one might consider the existential counterpart of this problem, i.e., whether there is some timed word $w \in T\Sigma^*$ such that $(\mathcal{T}_1, w) \leq (\mathcal{T}_2, w)$. A starting point for research on these problems could be the paper on quantitative languages by Henzinger et al. [38], where they have been investigated under the names of *quantitative universality* and *quantitative emptiness* problem. Also, Corollary 6.13 on the decidability of the equivalence of two recognizable timed series is a promising source of further resarch.

7 Conclusion and Future Work

In this thesis we studied characterizations of recognizable timed series. For this, we introduced a new definition of a weighted timed automaton over semirings and families of weight functions. We obtained a general model of weighted timed automata, in which we are not restricted to use a fixed set of weights and operations for computing the behaviour of the automaton. In particular, our model subsumes other weighted timed automata used in the literature so far, for instance the original proposals by Alur et al. [8] and Behrmann et al. [17] and timed automata with stopwatch observers [34]. We can also model new interesting instances of weighted timed automata never considered before, e.g. weighted timed automata over the Viterbi semiring and exponential functions to model probabilities. Our new definition discloses that this particular setting is isomorphic to the model of weighted timed automata over the min-plus-semiring and linear functions. Also, by defining weighted timed automata over a semiring, we build a bridge to the classical theory of weighted finite automata. In this thesis, we aimed to provide the reader with a new theoretical perspective on weighted timed automata by transferring some of the fundamental results from the theory of weighted finite automata to the timed setting. However, there is a big difference between weighted finite automata and weighted timed automata in that in the latter model one is allowed to assign weights not only to the edges (transitions) but also to the locations (states). This additional feature accounts for the practical interest in this model as it allows for the modelling of *continuous* resource consumption. However, it also complicates (or excludes, in the worst case) some of the constructions known from the theory of weighted finite automata. In the following, we summarize the main problems and their solutions.

First of all, we showed that due to the weight functions assigned to the locations, recognizable timed series are not closed under the Hadamard product in general. Recall that the Hadamard product corresponds to intersection if we consider weighted timed automata over the Boolean semiring and constant functions. We presented an example showing that the classical product construction does not work if the family of weight functions is not closed under the pointwise product. We defined a general condition on two weighted timed automata, captured by the notion of being *non-interfering*, for which the classical product construction works. However, notice that for weighted timed automata over the min-plus-semiring and the family of linear functions [8], we do not need to restrict the application of the Hadamard product to non-interfering weighted timed automata, since the family of linear functions is closed under the pointwise product.

Second, in chapters 4 and 5, we needed initial- and final-location-normalization constructions of weighted timed automata. These normalization techniques are well known from the theory of weighted finite automata [58]. However, the initial-location-normalization construction we propose here is a bit less restricted than that of Eilenberg. In his construction, there is a *single* location with an entering weight of 1, and the entering weights of all other locations are 0. Due to problems that arise with the location weight functions, we can neither construct an equivalent weighted timed automaton with a single location only, nor can we restrict the entering weights to 1. In the proof of the Kleene-Schützenberger theorem, we get by with this less restricted notion of initial-location-normalization. In chapter 5, however, we had to exclude the empty timed word in our construction of a logical sentence from a given weighted timed automaton due to this less restricted definition.

Third, in chapter 5, we extended the weighted MSO logic proposed by Droste and Gastin [46] with a new kind of weighted formulas to be able to define the weight functions assigned to the locations of a weighted timed automaton. We also had to extend the notion of almost unambiguous formulas to avoid the problems that occur in the combination of conjunction (whose semantics is defined by the Hadamard product) and these new atomic formulas. We point out that, unlike the untimed case [46, 49], the semantics of an almost unambiguous formula is not guaranteed to have a finite image. Nonetheless, we were able to adopt parts of the proof for closure under first-order universal quantification from Droste and Gastin [46]. A crucial part in the proof is a new normalization technique (Lemma 5.15). Furthermore, due to the non-determinizability of timed automata [6], when considering weighted timed automata and logics over non-idempotent semirings, we had to strike a completely new path. We restricted the application of the universal quantifiers of our logic to formulas whose semantics correspond to timed series over a restricted timed alphabet, namely a set of timed words with a bounded variability. For TA-recognizable timed languages over this restricted alphabet, a *deterministic* timed automaton is guaranteed to exist [106].

Fourth, the location weight functions may give rise to an infinite set of weights that occur in a weighted timed automaton. This is for instance the case for weighted timed automata over the family of linear functions. Here we cannot simply transfer constructions used for solving decision problems for supports and cut languages from weighted finite automata to weighted timed automata, as most of them rely on the finiteness of the number of weights occuring in the automaton, see e.g. the construction of a finite automaton recognizing the support of the behaviour of a weighted finite automaton over a commutative and zero-sum free semiring [75]. However, for some semirings and families of functions, we can show that the exact weights arising from letting time elapse in a location are not important to give a correct decision procedure. Amongst others, this is the case for the empty support problem for weighted timed automata over the semiring of the reals and the family of linear functions. Furthermore, when generalizing known results from the classical theory, we have to be a bit careful with zero time delays (or non-strictly monotonic timed series, respectively), e.g. in Theorem 6.2.

Overall, in this thesis we came up with two alternative characterizations of recognizable timed series, namely in terms of rational timed series and in terms of sentences of a weighted timed MSO logic. We presented the corresponding generalizations of two fundamental theorems in theoretical computer science, the Kleene theorem and the Büchi theorem. These results are not only of theoretical interest, but may be useful also for practical purposes. Rational expressions are widely used as they provide a concise formalism that can easily be read and interpreted by a rational expression processor. Also, the logical formula $\exists y.f(y)$ is much easier to read than the weighted timed automaton with the corresponding behaviour (see Fig. 5.1). Furthermore, the two main theorems in chapters 4 and 5 show the robustness of our model of a weighted timed automaton and recognizable timed series, respectively. We further like to mention that we can provide corresponding results for a strict subclass of weighted timed automata, called *weighted event-clock automata* [92, 94], a weighted extension of *event-clock automata* [7]. As a consequence of these results, the open problem whether we can provide Kleene and Büchi theorems for weighted timed automata over infinite timed words arises. For this, we have to consider convergency problems that occur when the weight of infinite runs is computed. However, for the untimed setting, there are several proposals to solve this problem [54, 55], which may be a promising starting point for research in this matter.

In future work, we would like to investigate whether there are timed extensions of *algebraic* characterizations of untimed series like linear representations [19] and aperiodic series [47]. In this context it may be useful to consider a weighted version of *data automata*, a strict superclass of timed automata introduced by Bouyer et al. [30], of which a monoid-based algebraic characterization (along the lines of the classical notion of monoid-recognizability) exists. In the theory of weighted finite automata, many important results rely on such algebraic characterizations, for instance, algorithms to minimize the size of a weighted finite automaton [19]. By adopting notions and concepts from the classical theory, we hope to obtain a deeper insight in our model and come up with new algorithms and instruments for the verification and analysis of real-time systems.

A first step in this direction was done in chapter 6, where we investigated the support and timed cut languages of recognizable timed series. Both notions are borrowed from the theory of weighted finite automata, where they have been investigated thoroughly [19, 98]. The results on supports we presented here shed light on the relation between weighted timed automata and timed automata. But first and foremost, they imply the decidability of weighted versions of some classical decision problems like the emptiness, universality and equivalence problem. We further investigated the TA-recognizability of timed cut languages of recognizable timed series. This may be useful from a practical point of view, when one is interested in checking whether the set of timed words which is assigned a weight greater than or equal to a given value satisfies a given property. We also argued that the results we presented here are connected to other interesting theoretical problems concerning e.g. unambiguously TA-recognizable timed languages.

An interesting future research question brought up by our new definition of a weighted timed automaton is whether recently obtained results concerning e.g. the optimal cost reachability problem [9, 17, 22], model checking [32, 21, 23, 34, 26] or weighted timed games [4, 33, 21, 23, 27] can be extended to our general model of weighted timed automata. We recall that these results only apply to weighted timed automata over the min-plus-semiring (max-plus-semiring, respectively) and the family of linear functions (and isomorphic structures, see Remark 6.20). Some of the results [17, 16, 9, 32, 21, 22, 26] are based on the classical region graph construction introduced by Alur and Dill [6] or extensions of it, respectively. In chapter 6, we defined a weighted extension of the region automaton to solve the empty support problem for weighted timed automata over the semiring of the reals and linear functions. It would be interesting to formulate general conditions on the semiring and the family of weight functions such that the behaviour of a weighted extension of the region automaton corresponds in some sense to the behaviour of the underlying weighted timed automaton. The results of this work may be significant for generalizing some of the results mentioned above on weighted timed automata. In the same spirit, we would like to investigate whether we can generalize the results on weighted timed games and model checking to arbitrary semirings and families of functions.

Furthermore, we want to move our research towards *quantitative languages*, which have been introduced by Henzinger et al. [38]. Quantitative languages are functions that map each word over an alphabet to a value. They can be recognized by quantitative automata, i.e., finite automata whose transitions are assigned weights, but, as opposed to weighted finite automata, these weights do not necessarily come from a semiring. Also the operations used to define the running weight and behaviour of such an automaton do not need to form a semiring. It is a natural step to lift the notion of quantitative languages to the timed setting. In fact, some of the previously defined models of weighted timed automata do not fit in our proposed model of weighted timed automata over a semiring, since the operations do not form a semiring (e.g. [80, 24]). We would like to define a more general notions of quantitative timed languages and quantitative timed automata, which allow for more general conclusions. Then it would also be interesting to establish the exact border between the two notions of recognizable timed series and quantitative timed languages, i.e., to figure out the similarities and differences between these two concepts.

Bibliography

[1] Y. Abdeddaïm, E. Asarin, and O. Maler. Scheduling with timed automata. *Theoretical Computer Science*, 354(2):272–300, 2006.

[2] Y. Abdeddaïm and O. Maler. Job-shop scheduling using timed automata. In G. Berry, H. Comon, and A. Finkel, editors, *CAV*, volume 2102 of *LNCS*, pages 478–492. Springer, 2001.

[3] Y. Abdeddaïm and O. Maler. Preemptive job-shop scheduling using stopwatch automata. In J.-P. Katoen and P. Stevens, editors, *TACAS*, volume 2280 of *LNCS*, pages 113–126. Springer, 2002.

[4] R. Alur, M. Bernadsky, and P. Madhusudan. Optimal reachability in weighted timed games. In J. Díaz, J. Karhumäki, A. Lepistö, and D. Sannella, editors, *ICALP*, volume 3142 of *LNCS*, pages 122–133. Springer, 2004.

[5] R. Alur, C. Courcoubetis, N. Halbwachs, T. A. Henzinger, P.-H. Ho, X. Nicollin, A. Olivero, J. Sifakis, and S. Yovine. The algorithmic analysis of hybrid systems. *Theoretical Computer Science*, 138(1):3–34, 1995.

[6] R. Alur and D. L. Dill. A theory of timed automata. *Theoretical Computer Science*, 126(2):183–235, 1994.

[7] R. Alur, L. Fix, and T. A. Henzinger. Event-clock automata: A determinizable class of timed automata. *Theoretical Computer Science*, 211(1-2):253–273, 1999.

[8] R. Alur, S. La Torre, and G. J. Pappas. Optimal paths in weighted timed automata. In M. D. Di Benedetto and A. Sangiovanni-Vincentelli [18], pages 49–62.

[9] R. Alur, S. La Torre, and G. J. Pappas. Optimal paths in weighted timed automata. *Theoretical Computer Science*, 318:297–322, 2004.

[10] R. Alur and P. Madhusudan. Decision problems for timed automata: A survey. In M. Bernardo and F. Corradini, editors, *SFM-RT*, volume 3185 of *LNCS*, pages 1–24. Springer, 2004.

[11] E. Asarin. Challenges in timed languages: From applied theory to basic theory. *The Bulletin of the EATCS*, 83:106–120, 2004.

[12] E. Asarin, P. Caspi, and O. Maler. A Kleene theorem for timed automata. In *LICS '97*, pages 160–171. IEEE Computer Society, 1997.

[13] E. Asarin, P. Caspi, and O. Maler. Timed regular expressions. *Journal of the ACM*, 49(2):172–206, 2002.

[14] E. Asarin and C. Dima. Balanced timed regular expressions. In W. Vogler and K.G. Larsen, editors, *MTCS*, volume 68 of *ENTCS*, pages 16–33. Elsevier Science Publishers, 2002.

[15] E. Asarin and O. Maler. As soon as possible: Time optimal control for timed automata. In F. W. Vaandrager and J. H. van Schuppen, editors, *HSCC*, volume 1569 of *LNCS*, pages 19–30. Springer, 1999.

[16] G. Behrmann and A. Fehnker. Efficient guiding towards cost-optimality in uppaal. In T. Margaria and W. Yi, editors, *TACAS*, volume 2031 of *LNCS*, pages 174–188. Springer, 2001.

[17] G. Behrmann, A. Fehnker, T. Hune, K. Larsen, P. Pettersson, J. Romijn, and F. Vaandrager. Minimum-cost reachability for priced timed automata. In M. D. Di Benedetto and A. Sangiovanni-Vincentelli [18], pages 147–161.

[18] M. D. Di Benedetto and A. Sangiovanni-Vincentelli, editors. *Hybrid Systems: Computation and Control, 4th International Workshop, HSCC 2001, Proceedings*, volume 2034 of *LNCS*. Springer, 2001.

[19] J. Berstel and C. Reutenauer. *Rational Series and their Languages.* Springer-Verlag New York, 1988. Current online version: http://www-igm.univ-mlv.fr/~berstel/LivreSeries/LivreSeries8march2009.pdf.

[20] B. Bollig and I. Meinecke. Weighted distributed systems and their logics. In S. N. Artëmov and A. Nerode, editors, *LFCS*, volume 4514 of *LNCS*, pages 54–68. Springer, 2007.

[21] P. Bouyer. Weighted timed automata: Model-checking and games. In S. Brookes and M. Mislove, editors, *MFPS'06*, volume 158 of *ENTCS*, pages 3–17. Elsevier Science Publishers, 2006.

[22] P. Bouyer, T. Brihaye, V. Bruyère, and J.-F. Raskin. On the optimal reachability problem on weighted timed automata. *Formal Methods in System Design*, 31(2):135–175, 2007.

[23] P. Bouyer, T. Brihaye, and N. Markey. Improved undecidability results on weighted timed automata. *Information Processing Letters*, 98(5):188–194, 2006.

[24] P. Bouyer, E. Brinksma, and K. G. Larsen. Optimal infinite scheduling for multi-priced timed automata. *Formal Methods in System Design*, 32(1):3–23, 2008.

[25] P. Bouyer, U. Fahrenberg, K. G. Larsen, N. Markey, and J. Srba. Infinite runs in weighted timed automata with energy constraints. In F. Cassez and C. Jard [37], pages 33–47.

[26] P. Bouyer, K. G. Larsen, and N. Markey. Model checking one-clock priced timed automata. *Logical Methods in Computer Science*, 4:1–28, 2008.

[27] P. Bouyer, K. G. Larsen, N. Markey, and J. I. Rasmussen. Almost optimal strategies in one clock priced timed games. In S. Arun-Kumar and N. Garg, editors, *FSTTCS*, volume 4337 of *LNCS*, pages 345–356. Springer, 2006.

[28] P. Bouyer and A. Petit. Decomposition and composition of timed automata. In J. Wiedermann, P. van Emde Boas, and M. Nielsen, editors, *ICALP*, volume 1644 of *LNCS*, pages 210–219. Springer, 1999.

[29] P. Bouyer and A. Petit. A Kleene/Büchi-like theorem for clock languages. *Journal of Automata, Languages and Combinatorics*, 7(2):167–186, 2001.

[30] P. Bouyer, A. Petit, and D. Thérien. An algebraic characterization of data and timed languages. In K.G. Larsen and M. Nielsen, editors, *CONCUR*, volume 2154 of *LNCS*, pages 248–261. Springer, 2001.

[31] M. Bozga, C. Daws, O. Maler, A. Olivero, S. Tripakis, and S. Yovine. Kronos: A model-checking tool for real-time systems. In A. J. Hu and M. Y. Vardi, editors, *CAV*, volume 1427 of *LNCS*, pages 546–550. Springer, 1998.

[32] T. Brihaye, V. Bruyère, and J.-F. Raskin. Model-checking for weighted timed automata. In Y. Lakhnech and S. Yovine, editors, *Proceedings of FORMATS-FTRTFT'04*, volume 3253 of *LNCS*, pages 277–292. Springer, 2004.

[33] T. Brihaye, V. Bruyère, and J.-F. Raskin. On optimal timed strategies. In Paul Pettersson and Wang Yi, editors, *FORMATS*, volume 3829 of *LNCS*, pages 49–64. Springer, 2005.

[34] T. Brihaye, V. Bruyère, and J.-F. Raskin. On model-checking timed automata with stopwatch observers. *Inf. Comput.*, 204(3):408–433, 2006.

[35] J. A. Brzozowski. Derivatives of regular expressions. *Journal of the ACM*, 11(4):481–494, 1964.

[36] J. R. Büchi. On a decision method in restricted second order arithmetics. In E. Nagel et al., editor, *Proc. Intern. Congress on Logic, Methodology and Philosophy of Sciences*, pages 1–11, Stanford, 1960. Stanford University Press.

[37] F. Cassez and C. Jard, editors. *Formal Modeling and Analysis of Timed Systems, 6th International Conference, FORMATS 2008, Saint Malo, France, 2008. Proceedings*, volume 5215 of *LNCS*. Springer, 2008.

[38] K. Chatterjee, L. Doyen, and T. A. Henzinger. Quantitative languages. In M. Kaminski and S. Martini, editors, *CSL*, volume 5213 of *LNCS*, pages 385–400. Springer, 2008.

[39] E. M. Clarke, O. Grumberg, and D. A. Peled. *Model Checking*. The MIT Press. Cambridge, Massachusetts. London, England., 1999.

[40] C. Cortes and M. Mohri. Context-free recognition with weighted automata. *Grammars*, 3(2/3):133–150, 2000.

[41] K. Culik and J. Kari. Image compression using weighted finite automata. *Comput. Graphics*, 17:305–313, 1993.

[42] L. de Alfaro, T. A. Henzinger, and R. Majumdar. Discounting the future in systems theory. In J. C. M. Baeten, J. K. Lenstra, J. Parrow, and G. J. Woeginger, editors, *ICALP*, volume 2719 of *LNCS*, pages 1022–1037. Springer, 2003.

[43] L. E. Dickson. Finiteness of the odd perfect and primitive abundant numbers with n distinct prime factors. *American Journal of Mathematics*, 35:413–422, 1913.

[44] V. Diekert and D. Nowotka, editors. *Developments in Language Theory, 13th International Conference, DLT 2009, Stuttgart, Germany, 2009. Proceedings*, volume 5583 of *LNCS*. Springer, 2009.

[45] H. Dierks. Finding optimal plans for domains with restricted continuous effects with uppaal cora : effects with uppaal cora. In S. Biundo, editor, *ICAPS*, pages 1–10. AAAI Press, 2005.

[46] M. Droste and P. Gastin. Weighted automata and weighted logics. In L. Caires, G. F. Italiano, L. Monteiro, C. Palamidessi, and M. Yung, editors, *ICALP*, volume 3580 of *LNCS*, pages 513–525. Springer, 2005.

[47] M. Droste and P. Gastin. On aperiodic and star-free formal power series in partially commuting variables. *Theory of Computing Systems*, 42(4):608–631, May 2007.

[48] M. Droste and P. Gastin. Weighted automata and weighted logics. *Theoretical Computer Science*, 380(1-2):69–86, 2007.

[49] M. Droste and P. Gastin. Weighted automata and weighted logics. In M. Droste et al. [50], pages 175–211.

[50] M. Droste, W. Kuich, and H. Vogler, editors. *Handbook of Weighted Automata*. EATCS Monographs in Theoretical Computer Science. Springer, 2009.

[51] M. Droste and D. Kuske. Skew and infinitary formal power series. *Theoretical Computer Science*, 366(3):199–227, 2006.

[52] M. Droste and U. Püschmann. On weighted Büchi automata with order-complete weights. *International Journal of Algebra and Computation*, 17(2):235–260, 2007.

[53] M. Droste and K. Quaas. A Kleene-Schützenberger theorem for weighted timed automata. In R. M. Amadio, editor, *FoSSaCS*, volume 4962 of *LNCS*, pages 142–156. Springer, 2008.

[54] M. Droste and G. Rahonis. Weighted automata and weighted logics on infinite words. In O. H. Ibarra and Z. Dang, editors, *Developments in Language Theory*, volume 4036 of *LNCS*, pages 49–58. Springer, 2006.

[55] M. Droste and G. Rahonis. Weighted automata and weighted logics with discounting. *Theoretical Computer Science*, 410(37):3481–3494, 2009.

[56] M. Droste, J. Sakarovitch, and H. Vogler. Weighted automata with discounting. *Inf. Process. Lett.*, 108(1):23–28, 2008.

[57] M. Droste and H. Vogler. Weighted tree automata and weighted logics. *Theoretical Computer Science*, 366(3):228–247, 2006.

[58] S. Eilenberg. *Automata, Languages and Machines*, volume A. Academic Press, 1974.

[59] Z. Ésik and W. Kuich. A semiring-semimodule generalization of ω-regular languages I. *Journal of Automata, Languages and Combinatorics*, 10(2/3):203–242, 2005.

[60] Z. Esik and W. Kuich. On iteration semiring-semimodule pairs. *Semigroup Forum*, 75(1):129–159, 2007.

[61] I. Fichtner. *Characterizations of recognizable picture series*. PhD thesis, Universität Leipzig, Institut für Informatik, Abteilung Automaten und Sprachen, 2006.

[62] I. Fichtner. Weighted picture automata and weighted logics. *Theory of Computing Systems*, 2010. in press.

[63] O. Finkel. Undecidable problems about timed automata. In E. Asarin and P. Bouyer, editors, *FORMATS*, volume 4202 of *LNCS*, pages 187–199. Springer, 2006.

[64] M. Fox and D. Long. Modelling mixed discrete-continuous domains for planning. *Journal of AI Research*, 27:235–297, 2006.

[65] M. Fränzle and M. Swaminathan. Revisiting decidability and optimum reachability for multi-priced timed automata. In J. Ouaknine and F. W. Vaandrager [90], pages 149–163.

[66] C. A. Furia and M. Rossi. MTL with bounded variability: Decidability and complexity. In F. Cassez and C. Jard [37], pages 109–123.

[67] U. Hafner. *Low bit-rate image and video coding with weighted finite automata.* PhD thesis, Universität Würzburg, Germany, 1999.

[68] K. Havelund, A. Skou, K. G. Larsen, and K. Lund. Formal modeling and analysis of an audio/video protocol: an industrial case study using uppaal. In *IEEE Real-Time Systems Symposium*, pages 2–13. IEEE Computer Society, 1997.

[69] F. Heidarian, M. Schuts, F. Vaandrager, and F. Zhu. Modelling clock synchronization in the chess gMAC WSN protocol. In *QFM*, 2009. To appear.

[70] T. Henzinger. The theory of hybrid automata. In *LICS '96*, pages 278–292. IEEE Computer Society Press, 1996.

[71] T. A. Henzinger, P.-H. Ho, and H. Wong-Toi. HYTECH: A model checker for hybrid systems. *International Journal on Software Tools for Technology Transfer*, 1(1–2):110–122, 1997.

[72] H. Kamp. *On tense logic and the theory of linear order.* PhD thesis, UCLA, 1968.

[73] F. Katritzke. *Refinements of data compression using weighted finite automata.* PhD thesis, University of Siegen, 2001.

[74] D. Kirsten. An algebraic characterization of semirings for which the support of every recognizable series is recognizable. In R. Královic and D. Niwinski, editors, *MFCS*, volume 5734 of *LNCS*, pages 489–500. Springer, 2009.

[75] D. Kirsten. The support of a recognizable series over a zero-sum free, commutative semiring is recognizable. In V. Diekert and D. Nowotka [44], pages 326–333.

[76] S. C. Kleene. Representation of events in nerve nets and finite automata. In *Automata Studies*, pages 3–42. Princeton University Press, Princeton, NJ, USA, 1956.

[77] W. Kuich and A. Salomaa. *Semirings, Automata, Languages*, volume 5 of *EATCS Monographs on Theoretical Computer Science.* Springer, Berlin, 1986.

[78] O. Kupferman and Y. Lustig. Lattice automata. In B. Cook and A. Podelski, editors, *VMCAI '07*, volume 4349 of *LNCS*, pages 199–213. Springer, 2007.

[79] K. G. Larsen, P. Pettersson, and W. Yi. UPPAAL in a nutshell. *International Journal on Software Tools for Technology Transfer*, 1(1-2):134–152, 1997.

[80] K. G. Larsen and J. I. Rasmussen. Optimal conditional reachability for multi-priced timed automata. In V. Sassone, editor, *FoSSaCS*, volume 3441 of *LNCS*, pages 234–249. Springer, 2005.

[81] K. G. Larsen and J. I. Rasmussen. Optimal reachability for multi-priced timed automata. *Theoretical Computer Science*, 390(2-3):197–213, 2008.

[82] O. Maler and A. Pnueli. On recognizable timed languages. In I. Walukiewicz, editor, *FoSSaCS*, volume 2987 of *LNCS*, pages 348–362. Springer, 2004.

[83] C. Mathissen. Definable transductions and weighted logics for texts. In T. Harju, J. Karhumäki, and A. Lepistö, editors, *Developments in Language Theory*, volume 4588 of *LNCS*, pages 324–336. Springer, 2007.

[84] C. Mathissen. Weighted logics for nested words and algebraic formal power series. In L. Aceto, I. Damgård, L. Ann Goldberg, M. M. Halldórsson, A. Ingólfsdóttir, and I. Walukiewicz, editors, *ICALP (2)*, volume 5126 of *LNCS*, pages 221–232. Springer, 2008.

[85] I. Mäurer. Weighted picture automata and weighted logics. In B. Durand and W. Thomas, editors, *STACS*, volume 3884 of *LNCS*, pages 313–324. Springer, 2006.

[86] I. Meinecke. Weighted logics for traces. In D. Grigoriev, J. Harrison, and E. A. Hirsch, editors, *CSR*, volume 3967 of *LNCS*, pages 235–246. Springer, 2006.

[87] M. Mohri, F. Pereira, and M. Riley. Weighted automata in text and speech processing. In András Kornai, editor, *Extended Finite State Models of Language*, pages 46–50, 1996.

[88] M. Mohri, F. C. N. Pereira, and M. Riley. Speech recognition with weighted finite-state transducers. In L. Rabiner and F. Juang, editors, *Handbook on Speech Processing and Speech Communication, Part E: Speech recognition*, volume to appear. Springer-Verlag, Heidelberg, Germany, 2007.

[89] P. Niebert, S. Tripakis, and S. Yovine. Minimum-time reachability for timed automata. In P. Groumpos, P. Antsaklis, and N. Koussoulas, editors, *MED*, 2000.

[90] J. Ouaknine and F. W. Vaandrager, editors. *Formal Modeling and Analysis of Timed Systems, 7th International Conference, FORMATS 2009, Budapest, Hungary, 2009. Proceedings*, volume 5813 of *LNCS*. Springer, 2009.

[91] J. Ouaknine and J. Worrell. On the language inclusion problem for timed automata: Closing a decidability gap. In *LICS*, pages 54–63. IEEE Computer Society, 2004.

[92] K. Quaas. A Kleene-Schützenberger theorem for weighted event-clock-automata. http://www.informatik.uni-leipzig.de/~quaas/weca_ks.pdf, 2008.

[93] K. Quaas. On the supports of recognizable timed series. In J. Ouaknine and F. W. Vaandrager [90], pages 243–257.

[94] K. Quaas. Weighted timed MSO logics. In V. Diekert and D. Nowotka [44], pages 419–430.

[95] R. Alur and T.A. Henzinger. Real-time logics: complexity and expressiveness. In *Fifth Annual IEEE Symposium on Logic in Computer Science*, pages 390–401, Washington, D.C., 1990. IEEE Computer Society Press, IEEE Computer Society Press.

[96] G. Rahonis. Fuzzy languages. In M. Droste et al. [50], pages 481–517.

[97] J. Illum Rasmussen, K. G. Larsen, and K. Subramani. Resource-optimal scheduling using priced timed automata. In Kurt Jensen and Andreas Podelski, editors, *TACAS*, volume 2988 of *LNCS*, pages 220–235. Springer, 2004.

[98] J. Sakarovitch. Rational and recognisable power series. In M. Droste et al. [50], pages 105–174.

[99] A. Salomaa and M. Soittola. *Automata-Theoretic Aspects of Formal Power Series*. Springer, New York, 1978.

[100] M. P. Schützenberger. On the definition of a family of automata. *Information and Control*, 4:245–270, 1961.

[101] M. P. Schützenberger. On finite monoids having only trivial subgroups. *Information and Control*, 8:190–194, 1965.

[102] W. Thomas. Automata on infinite objects. In J. van Leeuwen, editor, *Handbook of Theoretical Computer Science, Volume B: Formal Models and Sematics (B)*, pages 133–192. Elsevier and MIT Press, 1990.

[103] W. Thomas. Languages, automata and logic. In G. Rozenberg and A. Salomaa, editors, *Handbook of Formal Languages*, pages 389–485. Springer, 1997.

[104] S. La Torre, S. Mukhopadhyay, and A. Murano. Optimal-reachability and control for acyclic weighted timed automata. In R. A. Baeza-Yates, U. Montanari, and N. Santoro, editors, *TCS '02*, volume 223, pages 485–497. Kluwer Academic Press, 2002.

[105] H. Wang. On rational series and rational languages. *Theoretical Computer Science*, 205(1-2):329–336, 1998.

[106] T. Wilke. *Automaten und Logiken zur Beschreibung zeitabhängiger Systeme.* PhD thesis, Christian-Albrecht-Universität Kiel, 1994.

[107] T. Wilke. Specifying timed state sequences in powerful decidable logics and timed automata. In H. Langmaack, W.-P. de Roever, and J. Vytopil, editors, *Formal Techniques in Real-Time and Fault-Tolerant Systems*, volume 863 of *LNCS*, pages 694–715, Lübeck, Germany, 1994. Springer-Verlag.

Index

Symbols

1_L 10
$\mathbb{1}$ 16
$\langle\delta\rangle$ 8
$\lfloor\delta\rfloor$ 8
ε 5
ν 6
σ 53
ϕ 6
φ^+ 71
φ^- 71
φ_b 70
$\widetilde{\varphi}$ 64
$\varphi \xrightarrow{+} \psi$ 71
$\varphi \xleftrightarrow{+} \psi$ 71
Γ 5
Σ 5
$\Sigma_{\mathcal{V}}$ 53
Φ 6
$f_1 \odot f_2$ 16
$\langle f, k, a, \phi, \lambda\rangle$ 30
$|w|$ 5
$|w_a|$ 5
$\bar{y} \leq \bar{z}$ 84
$\lfloor\bar{y}\rfloor_z$ 84
$\|\mathcal{A}\|$ 12, 17, 29
$\|\mathcal{A}\|_{l,l'}$ 42
$\mathcal{E}_n$ 28
$\mathcal{I}$ 8
$\uplus$ 5
$\cong$ 8
$\odot$ 10
; 10
$\equiv$ 58
$\pm$ 85

A

$\mathsf{abs}(w)$ 8
assignment 53

B

Büchi Theorem 1, 3, 12, 51, 68, 105

C

$\mathcal{C}$ 6
$c_{\max}$ 8
$C_n\Sigma^+$ 28
$C_n\Sigma^*$ 28
c_x 8
Cauchy product 10, 29, 34
$\mathsf{cc}(I, x)$ 95
clock behaviour 29
clock constraints 6
clock semantics 28
clock series 27, 28
- proper, 29
- rational, 30
- recognizable, 29

clock valuation 6
clock variables 6
clock word 27, 28
- empty, 28

commute element-wise 10, 20, 76
concatenation 27, 28

D

definable 55
- $\overleftarrow{\mathcal{L}\mathsf{d}}(\Sigma)$, 53, 70
- $\mathsf{sR}\overleftarrow{\mathcal{L}\mathsf{d}}(\mathcal{K}, \Sigma, \mathcal{F})$, 66, 69

$\mathsf{sR}\overleftarrow{\mathcal{L}\mathsf{d}}^{\mathsf{b}}(\mathcal{K}, \Sigma, \mathcal{F})$, 75
$\mathsf{sR}\overleftarrow{\mathcal{L}\mathsf{d}}^{\mathsf{bnc}}(\mathcal{K}, \Sigma, \mathcal{F})$, 78
dg(V) 85
dom(w) 5

E
edge 6

F
$\mathcal{F}$ 16
family 16
field 9, 13, 88, 91
formal power series 10
formula
almost unambiguous, 57, 59, 69, 104
equivalent, 58
syntactically unambiguous, 71
weighted atomic, 54
Free(φ) 53
function
characteristic, 16
constant, 16
linear, 16, 19, 39, 83, 84, 89, 93, 94, 98
step, 16, 88, 93, 95, 96, 100

H
Hadamard product 10, 20, 76, 103

I
idempotent 9
I 95

K
k 10
$k\varepsilon$ 10, 29
K_A 10
$\mathcal{K}\langle\langle\Sigma^*\rangle\rangle$ 10
$\mathcal{K}\langle\langle C_n\Sigma^*\rangle\rangle$ 28
$\mathcal{K}^{\mathcal{F}-rec}\langle\langle C_n\Sigma^*\rangle\rangle$ 29
$\mathcal{K}^{\mathcal{F}-rec}\langle\langle T\Sigma^*\rangle\rangle$ 17
Kleene star 12, 29, 39
Kleene Theorem 1, 3, 48, 105
Kleene-Schützenberger Theorem . 46, 48

L
$L(\varphi)$ 53
$L_{\mathcal{V}}(\varphi)$ 53
$\overleftarrow{\mathcal{L}\mathsf{d}}(\Sigma)$ 52
$\overleftarrow{\mathcal{L}\mathsf{d}}(\mathcal{K}, \Sigma, \mathcal{F})$ 54
lab(e) 6
linear representation 13, 105
logic
relative distance, 52
weighted relative distance, 54

M
Min(M) 84
monoid 9
commutative, 9
morphism, 10
monomial 12, 29
$\mathcal{F}$-, 30
MSO($T\Sigma^*$) 52
MSO($\mathcal{K}, T\Sigma^*, \mathcal{F}$) 54

N
$\mathbb{N}$ 5
$N_{\mathcal{V}}$ 53
non-interfering 76, 103
normalized
final-location-, 32, 35, 39, 42, 96
initial-location-, 38, 39, 42

P
pointwise product 16, 103
problem
emptiness, 7, 13, 81, 87, 105
empty support, 13, 87, 91
equivalence, 105
language equivalence, 8
language inclusion, 8, 102
optimal weight reachability, 106
universal support, 91

universality, 8, 13, 87, 105
zero generation, 85

Q
$\mathbb{Q}$ 5

R
$\mathbb{R}$ 5
rwt 12, 17
recognizable
$\mathcal{F}$-, 17
deterministically TA-, 7
strictly monotonic $\mathcal{F}$-, 82
TA-, 6
unambiguously $\mathcal{F}$-, 94, 96, 98
unambiguously TA-, 7, 26, 92, 96, 98
WFA-, 12
region automaton 8
relative distance logic 52
relative distance predicate 52
renaming 6, 10, 24
inverse, 10, 25
ring 9
run 6, 11
running weight 11, 17

S
S 10
S^k 29
S^* 29
$\mathsf{sR}\overleftarrow{\mathcal{L}\mathsf{d}}^{\mathsf{bnc}}(\mathcal{K}, \Sigma, \mathcal{F})$ 77
scalar product 20, 24
Schützenberger Theorem 12, 27
semiring 9
Boolean, 9, 103
characteristic zero, 9, 92, 93
idempotent, 4, 9, 26, 56, 66, 68
locally finite, 9
max-plus-, 9, 84–86, 94, 97
min-max-, 9, 84–86, 97, 98, 100
min-plus-, 9, 19, 21, 83–86, 93–95, 100
morphism, 10, 95, 97
positive, 9, 82
Viterbi, 9, 95
zero-divisor-free, 9
zero-sum-free, 9
series 10
characteristic, 10, 26, 92
$\mathsf{source}(e)$ 6
stopwatch 19
sum 10, 20, 29
$\mathsf{supp}(S)$ 10
support 10, 12, 81, 82, 105

T
$\mathcal{T}$ 17
$T\Sigma^+$ 5
$T\Sigma^*$ 5
$T_b\Sigma^*$ 70
$T_s\Sigma^*$ 5
timed automaton 6, 18
deterministic, 7, 104
dual-priced, 19
multi-priced, 19, 83
stopwatch observer, 19
unambiguous, 7
underlying, 18
timed cut language 81, 92
timed language 5
cut, 92, 105
rational, 27
timed series 17
non-interfering, 21, 22
rational, 27, 47
strictly monotonic, 18, 84
timed word 5
strictly monotonic, 5
transition 6
discrete, 6
clock 28

timed, 6
clock 28

V
variability 70
bounded, 69, 104
$\mathcal{V}$ 53

W
$\mathsf{wgt}(\mathcal{A})$ 18
$\mathsf{wgt}_E(\mathcal{A})$ 18
$\mathsf{wgt}_{\mathcal{F}}(\mathcal{A})$ 18
$\mathsf{wgt}(\varphi)$ 76
$\mathsf{wgt}_E(\varphi)$ 76
$\mathsf{wgt}_{\mathcal{F}}(\varphi)$ 76
weighted finite automaton 11, 19
weighted timed automaton 15, 16
non-interfering, 21
unambiguous, 18

Z
$\mathbb{Z}$ 5
ZGP 85, 86

List of Figures

2.1 A timed automaton with a single clock 5
2.2 A weighted finite automaton over the semiring of the real numbers 11

3.1 A weighted timed automaton as defined by Alur et al. [8] 15
3.2 Weighted timed automaton for Example 3.1 18
3.3 Weighted timed automaton for Example 3.2 18

4.1 The weighted timed automaton for Example 4.5 31

5.1 $\mathcal{A}_{\overleftarrow{d}(D,y)\sim c}$, $\mathcal{A}_{\neg\overleftarrow{d}(D,y)\sim c}$, $\mathcal{A}_k$ and $\mathcal{A}_{f(y)}$ 60

6.1 A weighted timed automaton $\mathcal{A}$ with $\|\mathcal{A}\|_{=}^{-1}(6)$ not TA-recognizable . . . 94

List of Tables

5.1 Overview of Weighted Timed MSO Logics 78

Wissenschaftlicher Werdegang

Sep. 1992 – Jul. 1998	*Felix Klein Gymnasium Leipzig*
03.07.1998	Abitur
Okt. 1998 – Okt. 2002	*Hochschule für Technik, Wirtschaft und Kultur (FH) Leipzig* Studium Medientechnik Diplomarbeit "Suchmaschinen und Semantic Web. Neue Wege der Informationsbeschaffung", Betreuer: Prof. Dr.-Ing. Jörg Bleymehl, HTWK Leipzig
16.10.2002	Diplom Ing. (FH) "sehr gut"
Okt. 2002 – Aug. 2005	*Universität Leipzig* Studium Informatik
Aug. 2005 – Okt. 2006	*Universität Uppsala, Schweden* Studium Computer Science, Masterarbeit "On the Universality Problem for Timed Automata", Betreuer: Prof. Dr. Parosh A. Abdulla, Universität Uppsala
27.10.2006	Master of Science
seit Okt. 2006	*Universität Leipzig* Promotionsstudium am Institut für Informatik, Abteilung Automaten und Sprachen, Stipendiatin des DFG Graduiertenkollegs 446/3 "Wissensrepräsentation" sowie des Freistaates Sachsen (Landesgraduiertenstipendium), Betreuer: Prof. Dr. Manfred Droste, Universität Leipzig

Bibliographische Daten

Begutachtete Veröffentlichungen

1. On the Supports of Recognizable Timed Series. In *Formal Modelling and Analysis of Timed Systems 2009* (FORMATS 2009), Proceedings. Lecture Notes in Computer Science, Volume 5813, S. 243-257, Springer Berlin/Heidelberg. 2009.

2. Weighted Timed MSO Logics. In *Developments in Language Theory 2009* (DLT 2009), Proceedings. Lecture Notes in Computer Science, Volume 5583, S. 419-430. Springer Berlin/Heidelberg. 2009.

3. A Kleene-Schützenberger Theorem for Weighted Timed Automata (mit M. Droste). In *Foundations of Software Science and Computation Structures 2008* (FoSSaCS 2008), Proceedings. Lecture Notes in Computer Science, Volume 4962, S. 142-156. Springer Berlin/Heidelberg. 2008.

Eingereichte Manuskripte

1. MSO Logics for Weighted Timed Systems. Eingereicht bei *Formal Methods in System Design*. Juli 2009.

2. A Kleene-Schützenberger Theorem for Weighted Timed Automata (mit M. Droste). Eingereicht bei *Theoretical Computer Science*. September 2009.

Selbständigkeitserklärung

Hiermit erkläre ich, die vorliegende Dissertation selbständig und ohne unzulässige fremde Hilfe angefertigt zu haben. Ich habe keine anderen als die angeführten Quellen und Hilfsmittel benutzt und sämtliche Textstellen, die wörtlich oder sinngemäß aus veröffentlichten oder unveröffentlichten Schriften entnommen wurden, und alle Angaben, die auf mündlichen Auskünften beruhen, als solche kenntlich gemacht. Ebenfalls sind alle von anderen Personen bereitgestellten Materialien oder erbrachten Dienstleistungen als solche gekennzeichnet. Leipzig, 24. November 2009.

Karin Quaas